Environmental Evolution

Effects of the Origin and Evolution of Life on Planet Earth

second edition

edited by Lynn Margulis,
Clifford Matthews, and
Aaron Haselton

The MIT Press
Cambridge, Massachusetts
London, England

Set in Palatino by Wellington Graphics.
Printed and bound in the United States of America.

Library of Congress Cataloging-in-Publication Data

Environmental evolution : effects of the origin and evolution of life on planet earth.—2nd ed. / edited by Lynn Margulis, Clifford Matthews, and Aaron Haselton.
p. cm.
Includes bibliographical references and index.
ISBN 0-262-13366-0 (cloth :alk. paper) ISBN 0-262-63197-0 (pbk. : alk. paper)
1. Evolution (Biology) 2. Evolution. 3. Biosphere. I. Margulis, Lynn, 1938– II. Matthews, Clifford. III. Haselton, Aaron.

QH366.2.E59 2000
576.8—dc21

99-046389

front cover:
The surface waters of Lake Cisó in northeast Spain sometimes turn bright red when winds abate and temperatures rise. Ricardo Guerrero, who took the photographs on the front and back covers of this book, has shown that the color is due to great population densities of *Chromatium* and other anoxygenic photosynthetic bacteria. Today, *Chromatium* and other purple and green bacteria that use sulfide rather than water in their photosynthesis (and therefore deposit sulfur rather than expel oxygen) still generate this scene reminiscent of the Archean eon more than 3 billion years ago. At certain seasons, *Chromatium* and other purple photosynthetic bacteria descend to thrive out of the way of the oxygen, where there is plenty of sulfur just a few centimeters below the surface of the water; for that reason, we do not notice them. The background (drawn by Christie Lyons) suggests the complexity of Lake Cisó's *Chromatium* community.

back cover:
The surface waters of Lake Cisó usually lack the red color depicted on the front cover. The great population densities of *Chromatium* and other anoxygenic photosynthetic bacteria lie out of view below the surface. Above them, cyanobacteria, green algae, and other protists produce and remove oxygen. This scene is reminiscent of the Proterozoic eon, which began 2.5 billion years ago. The background (drawn by Christie Lyons) suggests the complexity of Lake Cisó's *Chromatium* community.

Dedicated to the memory of our dear colleagues Elso S. Barghoorn (1914–1984), Tony Swain (1922–1987), and Cyril Ponnamperuma (1923–1994).

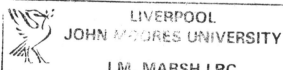

Contents

Foreword

In the first edition of *Environmental Evolution,* I paid tribute to the emerging field of planetary biology. I indicated that our previously insular approach to the biological specialties was being replaced by more global views of biogeochemistry and terrestrial ecology. At the time, I could see the speed of this change in our view of biology, but I could not sense the acceleration or the fiscal drivers.

We live at a time when change is in the air. Political, social, and economic change power the transition into the new millennium. The enlarged enrollments in the life sciences in higher education and the enormous interest in scientific literature about esoteric subjects such as the ozone hole, climate change, deforestation, and biodiversity are the heralds of a new period.

The demand for this second edition of *Environmental Evolution* is evidence for the hunger of our society to understand the Earth's biota and its relation to the ongoing processes. The editors of the first edition, Lynn Margulis and Lorraine Olendzenski, presented views way beyond their time. When Dr. James E. Lovelock and Professor Margulis published the Gaia theory in 1974, they could not have foreseen the extraordinary panorama of interdependence between the living and the nonliving worlds.

Margulis's familiarity with the deepest thinkers of our time, matched with her zest, energy, and urgency to "keep current," is miraculous. She has selected the very best authors to tell the story. This new edition is not only a key reference; it also is prepared to serve as a textbook for many new courses in planetary and global biogeochemistry that are popping up like dandelions in the spring.

NASA has always been proud of its association with Professor Margulis. In the first edition of this book, I made comments about the connection of it and its accompanying environmental evolution course to a new initiative, the Mission to Planet Earth (MTPE). MTPE is now a well-established program that includes scientific projects such as the Tropical Rainfall Measuring Mission (which measures global rainfall), Landsat (which provides detailed images of the Earth), and Terra (which will send data back as though it came from a firehose).

Now another new inspiration has appeared on the horizon. Astrobiology, the study of life in the universe, covers a continuous spectrum of theory and data from the "Big Bang" to the space age. We see a need to conjoin our understanding of the prebiological world of organic compounds, of Darwinian chemistry, of the RNA world of biogenesis with with our understanding of early evolutionary biology, of the current interactive complex living world, and of the evolving holistic planet. Pictures from the Hubble telescope have opened up new cosmological inquiry into the origin of the universe. In the same way, the recent discovery of biota at the bottom of the darkest regions of the ocean and of the extraordinary ability of organisms to live in salt crystals, in hot springs, or at extreme pH has inspired us to ask not only "How did life begin on Earth?" but also "How does life begin?" This new edition provides students, teachers, and other interested readers with a background for the real question: Is life a cosmic imperative?

Gerald A. Soffen
Director of University Programs
NASA Goddard Space Flight Center
Greenbelt, Maryland

Preface

This book arose out of Environmental Evolution, a one-semester course in which seniors and graduate students listen to and watch audiotaped lectures accompanied by electrowriter materials and slides. The "interactive lectures" are delivered by eminent researchers exploring the effects of the evolution of life on Planet Earth. The course, developed from 1972 through 1989 at Boston University, has been expanded to the biology and geology departments at the University of Massachusetts at Amherst, where a facility is dedicated to it.

The book—a study of the history of the environment from prebiotic times to the present—focuses on how the origin and the evolution of life have affected the surface of the Earth. The book, the lecture tapes, and scientific papers answer the need for an interdisciplinary overview. Aside from serving as the main text for a course in environmental science, the book is envisioned as a secondary text in biology, chemistry, and geology courses.

Scientific "facts" are always changing; the ways in which students create useful models of Earth's past environments change less rapidly. Because the body of material to be covered is potentially unlimited in detail and complexity, we present a series of essential concepts requisite to the reconstruction of the history of life on Earth from clues taken to be representational. The text invites students and professors to join the scientists in discovering the meandering paths of the evolution of life and the environment.

After an introductory discussion of the Gaia concept, the chapters are presented roughly in the chronological order of their subject matter, beginning with the origins of life. Each chapter opens with an abstract and is followed by a short list of recommended readings. Wherever possible, the author has revised the transcript to incorporate answers to questions asked by students and faculty members. (Professor Swain died without revising his chapter, so the questions—presented in their original form at the end of the chapter—were answered by his colleague Robert Buchsbaum. A few of the questions asked of Dr. Lovelock are presented separately because they concern his thoughts about the Gaia Hypothesis more than a decade after the original lecture was recorded.)

In the opening chapter, James Lovelock envisions Earth as a "blue marble" that regulates its surface far differently than would a planet similar in size and in position relative to the sun but devoid of life. In this early lecture, and in his responses to many questions raised by students and colleagues since he first came forth with his new idea about the "environment" as a part of the system itself, we learn from Lovelock about the development of his Gaia Hypothesis.

Michael McElroy presents a picture of Earth as a "terrestrial" or inner planet before life emerged. Clifford Matthews, David Deamer, and Antonio Lazcano introduce us to the problems of reconstructing the origin of life. They propose provocative experimental approaches for the study of prebiotic chemistry. Paul Strother and Elso Barghoorn describe attempts to establish the antiquity of life through the use of fossil evidence, warning us of the temptation to overinterpret the "organized entities" we encounter in the early fossil record. Stjepko Golubic explores microbial mats, the ancient benthic communities that bear witness to some of the earliest cosmopolitan, stable forms of life. Lynn Margulis describes the symbiotic origins of protoctist, animal, fungal, and plant cells and the peculiar sexual-motility systems of the eukaryotic microbial ancestors before global expansion. Mark McMenamin explores the mystery of the first appearance (600 million years ago) of complex life in the form of Ediacarans and animals.

Raymond Siever describes the discoveries that led to the revolution in geology, in which the theory of plate tectonics and continental drift replaced a myriad of ad hoc geological concepts. Tony Swain and Robert Buchsbaum introduce us to the chemical deterrents, warnings, and punishments that healthy, intact plants communicate to animals that would graze on them. Neil Todd offers a unique explanation of relatively recent mammalian evolution. Looking at the "adaptive radiations" of certain groups of mammals, he attempts to explain the episodes in which many new species of carnivores and artiodactyls appeared—evolutionary changes that are preserved in the mammalian fossil record. Todd correlates the changes in chromosome number that resulted from a process called karyotypic fission with these discontinuous episodes of evolution in representative mammalian groups since the beginning of the Cenozoic era. Finally, Jonathan King presents an analysis of environmental diseases brought about by industry's dismissal of natural history.

This book is dedicated to three colleagues who did not live to see it completed.

Elso S. Barghoorn, professor of paleobotany at Harvard University, played a crucial role in developing the idea of environmental evolution. A teacher as well as a profound scholar and thinker, he, more than any other biologist of the twentieth century, brought about an awareness of the immense amount of biological evolution that preceded the appearance of skeletalized animals. Barghoorn has been called the father of Precambrian paleobiology. As professor, colleague, and friend, he deeply influenced the development of the Environmental Evolution course as well as the careers and the thinking of all of us, especially Stjepko Golubic, Cyril Ponnamperuma, Lynn Margulis, and Paul Strother.

Tony Swain, during his tenure at the Royal Botanical Gardens at Kew and as a professor of biology at Boston University, investigated the evolution of communication between plants and animals. As cofounder (with Lynn Margulis) of the Planetary Biology Internship, which enables advanced students to participate in NASA's research activities in the life sciences, Swain was involved in the Environmental Evolution course as a classroom teacher. He was an inspirational force in the development of the program from 1979 until his death in 1987.

Cyril Ponnamperuma contributed a chapter on cosmochemical evolution to the first edition of this book, some of which has been incorporated here in Clifford Matthews's chapter. Cyril was the key worldwide figure in origin-of-life studies for most of his career, which began with his research in NASA's Exobiology Program. From 1971 to 1994 his research continued with the collaboration of visiting scientists, postdocs, and graduate and undergraduate students, primarily at his Laboratory for Chemical Evolution at the University of Maryland in College Park. Many of us first met one another at one of the regular interdisciplinary gatherings he held there. Cyril was one of the founders of the International Society for the Study of the Origin of Life (ISSOL), serving as editor of its journal for more than ten years. (Today the journal is called *Origins of Life and Evolution of the Biosphere*.) Cyril was one of the great champions of international efforts in science, with wide experience as technology advisor to several presidents of Sri Lanka, as founder of the Institute of Fundamental Studies in that developing country, and as President of the Third World Foundation of North America. In a memorial radio broadcast in Sri Lanka soon after Cyril's death in 1994, Arthur C. Clarke ended his eulogy with these words: "Hundreds of people of many nations—by no means all of them scientists—will miss his warm and compassionate personality." That includes many of us.

Acknowledgments

The Environmental Evolution course, its lectures and tapes (both audio and video), its organization, its field trips, and especially this text owe an immense debt of gratitude to the co-editor of the book's first edition. When Lorraine Olendzenski, now a research evolutionist and a superb science educator at the University of Connecticut in Storrs, found her current commitments overwhelming, course veteran Aaron Haselton (currently of the department of Entomology at the University of Massachusetts in Amherst) assumed Lorraine's duties in the preparation of this second edition.

We are grateful to Frank Urbanowski and Barry Silverstein, whose generosity permitted the transformation of these materials into a book. We applaud our authors for their unstinting cooperation and participation in this work, some for more than twenty years.

The book could not have been completed without the dedicated help of Jon Ashen, Paul Bethge, Christopher Brown, Lois Brynes, Emily Case, Michael Chapman, Margery Coombs, Eileen Crist, David Deamer, Kathryn Delisle, Michael Dolan, Betsey Dyer, Matthew Farmer, William Feder, Stephanie Hiebert, Gregory Hinkle, H. D. Holland, William Horgan, Jeremy Jorgensen, Margery Coombs, Wolfgang Krumbein, Heinz Lowenstam, Kelly McKinney, Hannah Melnitsky, Laura Nault, Karen Nelson, Donna Reppard, Dorion Sagan, Joseph Scamardella, Richard E. Schultes, David K. Scott, Mónica Solé, John Stolz, Sidney Tamm, Maud Walsh, Marta Norman, Madeline Sunley, and especially our MIT Press editor Clay Morgan. We also thank Frank Antonelli, Christopher Baldwin, Daniel Botkin, Fred Byron, Gillian Cooper-Driver, Beth Dichter, Michael Enzien, Amanda Ferro, René Fester, Gail Fleischaker, Ricardo Guerrero, William Irvine, George P. Fulton, Steven Goodwin, David Gorrill, Kate Gyllensvard,

Michael Keston, Alice (Sally) Klingener, Robin Kolnicki, Andrew Knoll, Thomas Kunz, Thomas Lang, Adam MacConnell, Sheila Manion-Artz, Heather McKhann, Donna Mehos, Bruce Parkhurst, Duncan Phillips, Mitchell Rambler, James G. Schaadt, Jacob Seeler, Linda Slakey, Valerie Vaughn, Constanza Villalba, Andrew Wier, and James Walker.

We acknowledge with gratitude the inventor of the interactive lecture system, Dr. Stewart Wilson, who has helped us in the many years we have used his system. The Polaroid Corporation supported both Wilson's work and some of ours by the donation of equipment.

Students at Boston University and at the University of Massachusetts and staff members of the Geddes Language Center at BU provided a continuous flow of aid and criticism; without their enthusiastic cooperation, we would have no book today.

Our greatest debt is to the staff of the NASA Life Sciences Office in Washington, Melvin Averner, Donald DeVincenzi, Arnauld Nicogossian, John Rummel, Michael A. Meyer, Gerald Soffen, and the late Richard Young. These colleagues are among those who have recognized the importance of the emerging science of global ecology in a context of comparative planetology. Since 1970 they have supported the unique research that generated the findings that have made this book possible.

Environmental Evolution

1   The Gaia Hypothesis

James E. Lovelock

When James Lovelock was first working out his Gaia concept of the physicochemical regulation of the Earth's surface, he visited the Environmental Evolution class several times. The disparate data and observations that led him to suggest a new view of life and the environment are revealed in this early lecture (from 1973) and in his responses to questions posed more than ten years later. We have presented them here both for historical interest and to illustrate the process of the development of scientific thought.

What has come to be called the Gaian view considers the atmosphere to be an integral part of the biosphere; the atmosphere here is not just a separate physicochemical system which interacts passively with life on Earth. Many will regard this as mere speculation, but I will try to prove that it is not. Even if I fail in this attempt, I think you will find that the Gaian view elicits new questions which otherwise might never have been asked.

Where are we? The small thatched building depicted in the foreground of figure 1 is my former laboratory at Bowerchalke in South Wiltshire. Built more than 400 years ago, it is, as far as I know, the only thatched space laboratory in the world. I am not introducing this territory as a bit of cozy folk science; the environment, as always, entails the organism.

The ground on which this laboratory was built is a well-known inorganic chemical substance calcium carbonate, or limestone: $CaCO_3$. The air—which, as you can see, is quite clear—is made up of oxygen (O_2), nitrogen (N_2), carbon dioxide (CO_2), water vapor (H_2O), and other inorganic chemicals. The plant segment of the biosphere colors the biosphere green. The animals are not so conspicuous, but they exist.

Figure 1
The author's laboratory in southwest England.

Life amid inanimate surroundings, life on its inert environment: this is the ordinary, traditional view of the biosphere. By contrast, the picture I'm developing is that the air and the ground are not independent inorganic chemicals, but that the sediments and atmosphere are part of a living system. On the other hand, despite its seeming position as part of pristine, unspoilt nature, even the green that surrounds my laboratory is not "natural": it is nearly all man-made, biology wittingly or unwittingly ordered by man over the course of time. Without people, this part of southern England would probably be primeval scrub forest.

From this point of view, air pollution on a global scale might perturb not just the atmosphere but the biosphere. The possibility of air pollution on a global scale began to attract professional interest in the early 1970s. My interest was sparked by the reports *Man's Impact on the Global Environment* and *Inadvertent Climate Modification*, published by The MIT Press in 1970 and 1971. When I read them, it occurred to me that a special view of the Earth was denied to the distinguished groups that produced the reports, a view that was more consonant with those of my colleagues working in the planetary sciences. This special view came from a need to look at the planet in its entirety, and in an interdisciplinary manner. By contrast, the view of these otherwise wholly excellent books is limited by the division of science into arbitrary disciplines. For example, the meteorologists state explicitly that they do not at all consider the chemistry or the biology of the Earth. The atmospheric chemists, for their part, say that meteorology lies beyond their territory; they make no reference to biology. This recognition of territorial rights is instinctive to most male animals, and this includes scientific experts and university professors especially. Such a comment may seem glib, yet it accurately accounts not only for the limitations of the MIT reports but also for those of current science in spite of the growing interest in and sympathy toward more interdisciplinary approaches.

Earth's Atmosphere: Evidence for Life

The problem of detecting the presence (or, much more likely, the absence) of life on other planets demands a less divided view. The search for life elsewhere brings biologists and engineers, for example, together in constructive conversations. For some years I have, with colleagues such as these, been interested in the possibility of detecting the presence of life on other planets merely from the knowledge of the chemical composition of the planet's atmosphere. I hypothesized that Mars would be without nitrogen, since it was probably without life; at least, this is what we surmised after seeing the cratered, moon-like surface revealed by the first Mariner mission. To understand the reasons for this educated guess, let us consider the relationship of nitrogen in the atmosphere to the presence or absence of life on Earth.

The cosmic abundance of elements is fairly constant, and (apart from an absence of hydrogen, which may have escaped to space) Earth is fairly representative of the general distribution of elements. These elements tend to combine to the state at which the lowest potential energy is reached; this is a law of chemistry. Comparison of the major constituents of Earth's atmosphere with those of Mars and Venus (table 1), however, reveals that Earth's atmosphere is anomalous with respect to these gases. With the chemical mixture present in the Earth's atmosphere, the element nitrogen is expected to form its most stable compound, which is not N_2 but the nitrate ion (NO_3^-). One would

Table 1
Major features of the planetary atmospheres compared: percent by weight of the reactive gases carbon dioxide, nitrogen, and oxygen; water in precipitable meters over the planet if all vapor precipitated out of the atmosphere; pressure; and mean annual surface temperature.

	Venus	Earth	Mars
CO_2	98%	0.03%	95%
N_2	1.7%	79%	2.7%
O_2	trace	21%	0.13%
H_2O	0.003 m	3,000 m	0.00001 m
Pressure	90 bars	1 bar	0.0064 bar
Temperature	477°C	17°C	−47°C

expect NO_3^- to be present either on the surface or in the seas as a potassium or sodium salt. Conversion of the nitrate ion to nitrogen gas is an "uphill" process which requires the presence of life. The expected chemical conversion of nitrate makes the presence of nearly 80 percent molecular nitrogen in Earth's atmosphere an indication of the presence of life.

The next Mariner mission found only 2.7 percent nitrogen in the Martian atmosphere. The large amount of nitrogen on Earth relative to Mars supported my view that life on Earth shows itself as a global chemical phenomenon.

When I was a young man, the atmosphere of Earth was said to have originated in the primeval outgassings from the Earth's interior; outgassing is a process, particularly important in the early history of the Earth, whereby gaseous and volatile compounds escape from the interior of the planet and help produce the early oceans and atmosphere.

Oxygen, for example, was explained to have come from the photodissociation of water vapor by sunlight followed by hydrogen escape. Nitrogen was said to have been a stable, inert constituent throughout Earth's history. Until a few years ago, this view was largely unchallenged. Indeed, it is widely held, and you will still find it in many textbooks on the Earth's atmosphere. In spite of the distinguished science done by such workers as L. V. Berkner and L. C. Marshall, who proposed a wholly biological origin for atmospheric oxygen, most aeronomists still believe that life, responsible for no net increment of oxygen, merely recycles oxygen gas. Oxygen, that is, just happens to be in the air at exactly the right concentration for most life because of blind inorganic processes. Or, if it is your preference, oxygen was arranged by some beneficent providence. To me this view of the air as a product of wholly inorganic processes is the most magnificent nonsense; indeed, one of the most intriguing puzzles in the history of science is how it has managed to persist for so long. G. N. Lewis and M. Randall in the 1920s, G. E. Hutchinson in the 1950s, and, most recently, L. G. Sillen all showed that at the pH and the redox potential of the Earth molecular nitrogen is thermodynamically unstable. The element nitrogen, given these conditions, should be present on Earth not as the gas but as nitrate ion in seas. From an inorganic viewpoint, the presence of oxygen is, in fact, equally anomalous when Earth is compared with its neighboring planets, Mars and Venus, which lack atmospheric oxygen. That Mars and Venus have no free oxygen is not unexpected. It is, rather, the presence of free, extremely reactive atmospheric oxygen on Earth that is a complete anomaly. The once-reducing atmosphere of the Earth is today an oxidizing one.

The opposed terms *reducing* and *oxidizing*, roughly equivalent to "hydrogen-rich" and "hydrogen-poor," come from old-fashioned chemistry. A reducing substance, such as hydrogen, will combine with the oxygen of a metal oxide, such as iron oxide (rust), to give the oxide of the reducer (in this case water) and also free iron as the metal. In a reducing atmosphere, iron would remain as metal. In an oxidizing atmosphere, it rusts—that is, iron recombines with oxygen to give its oxide again.

Another shibboleth which has held up progress in this branch of science is that the climate and the chemical composition of the Earth are uniquely favorable for life. This is the one you will find in most science fiction stories, sad to say. I say it is sad because, on the whole, science fiction tends to be a little less blind than science itself. Indeed, it is not commonly appreciated that seemingly quite small changes would render Earth unsuited to contemporary life.

Life Regulates the Environment

What would happen, for example, if oxygen were to increase in concentration? If it were just over 25 percent, the probability of a fire starting by a lightning flash would be so high that even tropical rain forests would be at risk. For each 1 percent increase in oxygen concentration over the current level, the probability of initiating combustion doubles. Interestingly, if the oxygen concentration were to fall to 13 percent one could not start any sort of fire at all. Nearly no difference exists in flammability between O_2 concentrations of 40 percent and 100 percent. Indeed, the current 21 percent O_2 concentration is just about ideal for the existence of trees, which participate in making oxygen: at higher levels trees would be burnt up; at lower ones there might well be too few animals and other oxygen consumers to maintain an ecological balance.

For another example of a quite small change that would have drastic consequences for present life forms, consider that a change in atmospheric pressure of merely 10 percent, assuming that the composition of air was unaltered, would result in a 4°C change in the world's mean surface temperature! Such a change would set Earth on a highly unfavorable climatic course. These examples, which show just how well suited the present atmosphere is to the present form of life on Earth, could be multiplied. I think the biota, the sum of all living organisms, interacts actively with its environment so as to maintain the environment at values of its own "choosing." The notion that blind chance led to such a perfectly adjusted atmosphere, by contrast, seems untenable.

Early in its evolution, life acquired the capacity, I believe, to control the global environment to suit its needs. The capacity for environmental maintenance has persisted; it is still active. The sum total of all the species that go to fill up the biosphere is far more than just a catalog; like other associations in biology, this global biota is an entity with properties greater than the sum of its parts. Such a large creature, with the powerful capacity to air-condition the whole planet, may be only hypothetical at the moment; nonetheless, it needs a name. I am grateful to Mr. William Golding, who lives in my village, for the suggestion of the word *Gaia:* the Greek personification of Mother Earth. It has various advantages, not least of which is its status as a four-letter word with the capacity to focus the attention of my scientific colleagues; certainly "Gaia" is a lot easier to say than "a biological cybernetic system with homeostatic tendencies."

Any theory stating that the Earth's surface is wholly a product of biological processes must be considered wrong. Comprehensive biological explanations of atmospheric composition are difficult to formulate. At the root of this problem is the fact that there exists no formal scientific statement of life as a process. I know of no single and exclusive test that could prove or disprove the existence of Gaia as a living entity. Biologists, fortunately, usually are not deterred by such a lack of rigor. Even if eventually prepared, such a formal physical statement of life is liable to be statistical, mechanical, very mathematical, and quite unsuited to the design of simple experiments to test for the presence of Gaia.

Most biologists, indeed most people, upon seeing a giraffe even for the first time, and especially if it moved, would be able to pronounce it alive without any conscious use of chemistry or physics. Life is still much in the realm of phenomenology. One scientific approach attractive to interdisciplinary biologists would attempt to prove the presence of life by seeing if the entity tested were able to maintain a constant temperature and compatible chemical composition in the face of environmental change or perturbation. From such a phenomenological basis, what evidence points to the existence of Gaia, a creature made up of the biosphere but more than just the sum of its parts?

We believe that during the period in which life has existed on Earth, from at least 300 million years ago to the present, the reduction-oxidation (redox) potential of the atmosphere has shifted from a pE of −5 (reducing) to a pE of +13 (oxidizing). (A useful way of indicating the oxidizing or reducing tendencies of a system, pE, the logarithm of the reciprocal of the electron concentration, is expressed in gram-molecules per liter.) During the time in which the pE changed from −5 to +13, the atmospheric composition changed concomitantly. The change from a reducing to an oxidizing atmosphere led to many correlated changes in atmospheric composition and total pressure. The early atmosphere was probably rich in hydrogen and ammonia. At the same time, as it moves along the standard course for average stars, the energy output of the sun has increased at least 30 percent. This increase in solar luminosity is one of the few relatively certain facts of astronomy. In spite of extensive atmospheric chemical changes and changes in output of radiant solar energy, the geological record, with its demonstrable persistence of life, indicates that at no time in the last 3 billion years did the Earth's mean temperature change more than a few degrees from what it is now. What sort of remarkable coincidence might account for such physical constancy, which is exactly what is required for the continued existence of life? Indeed, I am doubtful that this is coincidence; I think it very much more likely that a biological regulatory system has been and is working, ensuring planetary homeostasis at physical and chemical states appropriate to the global biota at a given time.

Life Regulates Global Mean Temperature

The most important evidence for Gaia is found in the constancy of the Earth's mean temperature through time. Since liquid water has always been present, the average temperature is unlikely ever to have exceeded 50°C or to have decreased much below the mean temperatures during the Pleistocene ice ages. Our sun, like all main-sequence stars, has been increasing in luminosity since its origin. Near the origin of the Earth and life, some 400 million years ago, the sun is thought to have been 30 percent fainter than at present. Quite clearly, even a comparatively short time ago, the Earth's temperature, if it responded passively to solar luminosity in the same atmosphere, would have been much lower than it is now. This paradox—the fact that the solar output was much weaker in the past, and yet Earth's mean temperature seems to have remained within certain boundaries—has been called "the Faint Young Sun Paradox."

To maintain a constant surface temperature, assuming current atmospheric composition, a decrease in atmospheric pressure would have to have occurred over time. On a mountain at a point where the pressure has dropped by 10 percent, you will find that the temperature has fallen by 4°C. Similarly, if you descend 1000 feet below sea level (as in a deep depression at the Dead Sea), you will find an equivalent rise in temperature. A 10 percent increase in atmospheric pressure corresponds to a rise in temperature of about 4°C. Meteorologists refer to this as the *adiabatic lapse rate*. Atmospheric pressure would have had to decrease to compensate for the increasing output of the sun with time.

What atmospheric composition provided higher atmospheric pressure in the past, leading to maintenance of constant temperature over geologic time? A greater amount of oxygen in the past is precluded.

Even small changes in oxygen levels would render disastrous effects. Drastic CO_2 rise is precluded by inorganic equilibria in water which hold this gas fairly near present level. The only gas that changes easily is nitrogen. About 1.5 billion years ago, about 20 percent more nitrogen may have been in the atmosphere than now. We can check this by seeing if younger rocks contain more nitrogen than older ones, since some of this extra nitrogen would have to have been buried to reduce the amount of nitrogen in the younger atmosphere. In fact, younger rocks do, I believe, by and large, contain more nitrogen than older ones. I view this as one possibility of "gaian" pressure and hence temperature regulation of the atmosphere, via manipulation of the quantity of atmospheric nitrogen.

But affairs are more critically balanced. If the mean surface temperature of Earth falls by more than 2 or 3°C, then positive feedback mechanisms associated with the increase of snow cover hasten the falling temperature, as is well established in meteorology. The Earth cools to a point where even the oceans may freeze. Earth's temperature is poised between hot and cold extremes. A 2 or 3°C rise would increase greenhouse gases in the atmosphere and, by a similar positive-feedback mechanism, cause further increased heating. With a sufficient rise in temperature, a thermal "runaway effect" would occur. Yet the temperature has remained relatively constant in the face of perturbation. To me this is the best evidence of the existence of Gaia: a system that controls the conditions at the surface of the planet.

If all of the life on Earth were deleted at one stroke, many inorganic atmospheric reactions—from electrical discharges and ionizing radiation to solar ultraviolet rays—would permit oxygen and nitrogen to react with one another. A predominant end product of these reactions of O_2 with N_2 would be the nitrate ion, which would wash into the sea. Nitrogen so removed would not return to the air: biological processes alone perform nitrate reduction to molecular nitrogen. Whereas the burying of nitrogen from the sea in plate-tectonic processes involves relatively slow recycling, we are concerned here with relatively short periods of about a million years. Small quantities of oxygen might be supplied by water or CO_2 photolysis (the breakup of carbon dioxide via lightning), ensuring the removal of the remaining nitrogen after the last of the oxygen has reacted. The end result of the steady-state inorganic equilibrium, in the absence of life, would be an atmosphere of CO_2, water vapor, carbon monoxide, and rare gases; only trace amounts of oxygen and nitrogen would persist. This scenario for the chemical composition of the atmosphere of a lifeless Earth is a very reasonable interpolation between the atmospheres of present-day Mars and Venus. The redox Earth, compared with its imaginary sterile twin, as well as oxidizing Venus and Mars on the one hand, and reducing Jupiter on the other, is portrayed in the bar graphs in figure 2.

Biotic Emissions Have Global Effects

Now let us consider: Do the activities of man have atmospheric effects, adverse or otherwise, on a global scale? The various gases of the atmosphere, their concentrations, their emission rates from the biosphere, their atmospheric residence times in years, and their emission rates from man-made sources are shown in table 2.

The first thing you will notice when you look at this table is the overwhelming dominance of the biota. The gases turn over quickly; most of them have short residence times. Oxygen, for example, although fully 21 percent of the total atmosphere, turns over once every thousand years. The residence time for N_2 is measured on the order of a million years—longer than other elements but still short compared to the time that life has been present on Earth. Carbon monoxide and oxides of nitrogen are produced not solely from the exhaust of cars but by the biota as well. Many creatures in the sea find carbon monoxide an essential part of their everyday business. Siphonophores float using little bladders filled with 80 percent carbon monoxide. Even *Fucus,* a common brown seaweed, has 0.1 percent carbon monoxide in its bladders. Carbon monoxide is not a noxious toxic emission, but part of the vital existence of these marine organisms. At least 1.2 billion tons of ammonia are annually produced by the biota. I say "at least" because the figures are probably underestimates; living organisms as new sources of gases are continually discovered. This very large production of ammonia has an important bearing on the maintenance for life of planetary pH. I think biogenic ammonia compensates for the tendency of the planet toward acidity. In regions such as northern Sweden, the northern United States, and Canada, ammonia production is apparently deficient; rain as acid as pH 3 sometimes falls.

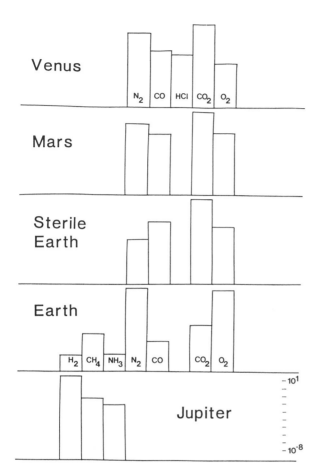

Figure 2
Concentrations of reactive gases in the atmospheres of the inner planets of the solar system on a log scale (from 10^{-9} to 10^1). From most reducing (left) to most oxidizing (right): hydrogen, methane, ammonia, nitrogen, carbon monoxide, hydrogen chloride, carbon dioxide, oxygen. "Sterile Earth" was calculated by interpolation between Mars and Venus or by letting "Earth" gases react with each other to chemical equilibrium.

Table 2
Atmospheric gases: sources, residence times, and primary sinks.

Gas	Formula	Components of air (ppm) by volume	Sources (atmospheric emissions) (billions of tons/year)			Residence time[f] (years)	Principal sinks[g]
			Abiological Volcanic & tectonic emissions	Biological Microbes, plants, animals	Man-made		
Nitrogen	N_2	780,300	—[a]	1	0	500,000	Utilized by organisms
Oxygen	O_2	209,480	0.001	100	–10	6,000	Utilized by organisms, combusted
Carbon dioxide	CO_2	350	0.01	140	16	2–5	Utilized by organisms
Methane	CH_4	1.7	0	2	0	7–10	Utilized by organisms, atmospheric oxidation
Nitrous oxide	N_2O	0.31	—	1	0	150	Utilized by organisms
Carbon monoxide	CO	0.20 (NH)[b] 0.05 (SH)[c]	0.11	0.1	0.28	0.3	Utilized by organisms
Ammonia	NH_3	0.0001–0.001	0	1.2	0	1 week	Utilized by organisms
Hydrocarbons	$(CH_2)_n$	0.001	0	0.2	0.2	0.01–0.02	Atmospheric oxidation
Oxides of nitrogen	NO_x	$1-3 \times 10^{-5}$ [d] $1-2 \times 10^{-4}$ [e]	—	—	0.16	1–7 days	Utilized by organisms, atmospheric reactions
Hydrogen sulfide	H_2S	0.0001	—	0.1	0.003	0.005	Atmospheric oxidation
Sulfur dioxide	SO_2	0.00001– 0.0001	—	0	0.16	5 days	Atmospheric oxidation
Dimethylsulfide	$(CH_3)_2SCH_3$	0.0001	0	0.2	0	0.01	Atmospheric oxidation

Chlorofluorocarbons							
CFC-11	CFCL$_3$	0.0003	0	0	0.0003	75	??
CFC-12	CF$_2$CL$_2$	0.0004	0	0	0.0004	110	??
Sulfur hexafluoride	SF$_6$	2×10^{-13} ??	0	0	$+5 \times 10^{-7}$	30	None
Noble gases							
Argon	Ar	9,340					
Helium	He	5.2					
Neon	Ne	18.2	No organism interaction[h]				
Krypton	Kr	1.1					
Xenon	Xe	0.09					

a. Unknown.
b. Northern hemisphere.
c. Southern hemisphere.
d. Remote.
e. Populated.
f. Residence time, which can be thought of as approximately half-life, refers to the amount of time it takes for the values of these gases to fall to 1/e, or about 37 percent.
g. Fate of the gases as they are removed from the atmosphere.
h. Nongaian gases.

Methane is produced in quantities of more than 2 billion tons per annum. I believe the huge production of methane, representing some 8 percent of all of the energy of photosynthesis, is very important for the regulation of the atmosphere. One function of methane is the formation of a very convenient "molecular balloon"; this gas passes the atmosphere to regions where photolysis occurs and hydrogen escapes. Water vapor does not easily enter the upper atmosphere: the tropopause is very cold, and water vapor freezes out to a concentration of 0.5 ppm. Methane, in the same region, has a concentration of 1.5 ppm. Since each methane (CH_4) molecule carries twice as much hydrogen as water, six times as much hydrogen is carried up and outward by methane as by water. In addition, the Earth maintains a net oxidizing state by expelling hydrogen to space. If methane is involved in oxygen regulation, a mature ecosystem is justified in squandering as much as 8 percent of its energy to produce it.

The atmosphere is not a static mixture of gases preserved by Earth's gravitational field. It is a system in dynamic and contemporary balance (table 2). Undoubtedly the immediate origin of the atmosphere is the biota. I am certain that the relative constancy of atmospheric composition over time is actively maintained by sensing and control mechanisms within the biosphere. The human industrial sector, with the possible exception of carbon dioxide emission, contributes relatively little.

Can large-scale atmospheric effects, for example combustion emissions, be used as perturbations to test for Gaia? We first notice air pollution on a large scale by the presence of smoke haze, such as the atmospheric turbidity of a locale such as the British industrial city of Sheffield, a small steel town. By contrast, Bantry Bay in southwestern Ireland is far removed from industry and pollution sources. When the wind blows from the North Atlantic, the air is sparklingly clear. Visibility at ground level may be more than 40 miles. When the wind blows into Bantry Bay from continental Europe, a source of all sorts of pollution, the picture changes; the visibility range may be reduced to fewer than 1.5 miles. Alan Eggleton and his colleagues at Harwell Laboratory inform us that the rural summertime haze of northern Europe, like that at Sheffield, is a form of photochemical smog frequently associated, like Los Angeles smog, with comparatively high levels of ozone. When the wind blows from continental Europe, a level of about 0.1 part per million of ozone would be expected during the day, to be compared with about one-tenth of that in clear air conditions.

Atmospheric turbidity is a measure of the scattering of incoming sunlight by any particulate matter in the air. Dust particles are blown from the desert by wind or from farm lands under dust bowl conditions. But particulate matter, including fine droplets, may also be produced by reactions among the gases in the atmosphere. Sulfur compounds react with oxidants to produce sulfuric acid droplets or ammonium sulfate aerosols. Trees produce unsaturated hydrocarbons like pinene (a terpenoid), which polymerizes to form pinene polymer particles, an aerosol that scatters incoming sunlight. Dust particles can be generated in situ; they need not have been stirred up. Atmospheric turbidities plotted monthly at Bowerchalke (southern England), Greensborough (North Carolina), and San Bernardino (in the heart of the California smog basin) are indistinguishable. They all show the same high-level seasonal increase. Few people realize that the rural regions of southern England, the Appalachian Mountains, and Los Angeles all share the same density of turbid aerosol in the summertime. Though this turbid aerosol is of photochemical origin at all three sites, the haze at Bowerchalke and Greensborough is quite different from that of Los Angeles. At its worst in southern England or Greensborough, there is little or no odor or eye irritation; the aerosol appears to be principally composed of natural ammonium sulfate or sulfuric acid droplets.

A puzzling feature of smog in rural regions away from industry is that it is most marked with winds from directions which also lack industry. In southern England the densest smogs come with winds from a southeasterly quarter, which traverse many miles of open sea and rural areas before arriving. By contrast, air masses coming from the north over the densely populated industrial regions of the United Kingdom are comparatively free of this sort of haze. Seasonal change in atmospheric turbidity can be measured for air masses from three principal directions at Bowerchalke. Continental air masses have high levels of summertime haze. Air masses coming from a maritime tropical direction off the southern part of the Atlantic, or a maritime polar direction coming from the north, either over the sea or over the industrial regions of the United Kingdom, carry little smog. No significant smog occurs with winds from northerly directions! The densest turbidity in the United States, too, is found not in the densely populated industrial regions, but in the southern Appalachian Mountains, where urbanization is light but trees abound.

A compilation of the areas of turbidity in the northern hemisphere can be made from satellite photographs. The regions of densest turbid aerosol are not the industrial regions of western Europe, the United States, or Japan, but the tropical and desert regions of the world near the equator! The whole of Africa in its middle region lives in a state of permanent haze; this appears to be true also of much of southeast Asia. A sunset on the harbor at Dakar, Senegal, has that same golden look seen in Los Angeles. The Senegal haze, however, is not industrial, for Dakar is quite a small city with little industry; rather, it is the natural haze of Africa. The Canary Islands, photographed from a ship 5 miles offshore, can barely be seen, so dense is the turbidity. Whatever causes this atmospheric turbidity potentially affects the Earth's surface temperature in one way or another. Haze and atmospheric particulates come from the nonhuman biota as well as from industrial and domestic sources.

In the last century the concentration of atmospheric carbon dioxide has increased as a consequence of the ever-increasing burning of fossil fuels. It has risen from 280 parts per million to 350 parts per million, and the increase is beginning to accelerate. This increment corresponds to the retention in the atmosphere of about half of the carbon dioxide from fuel that has been burned. This increase anticipates a warming of the planet by the well-known greenhouse phenomenon: the absorption of infrared radiation by carbon dioxide, which lessens the heat loss of the Earth into space.

These atmospheric changes may be due largely to human activity, or maybe not. Changes in climate might be associated with these measured atmospheric changes. Since about 1925 the climate of the northern hemisphere has changed, and up until the winter of 1970–71 the change had continued in the same direction (figure 3). In the second of the MIT publications I mentioned (*Inadvertent Climate Modification*), a diagram shows the reduction in temperature in the northern hemisphere plotted against latitude for the period 1960–1965 and the period 1965–1971. A clear-cut, quite unequivocal decrease in temperature occurred for all regions higher than about latitude 50° north. A slight warming trend may have occurred near the equator, but it is nowhere

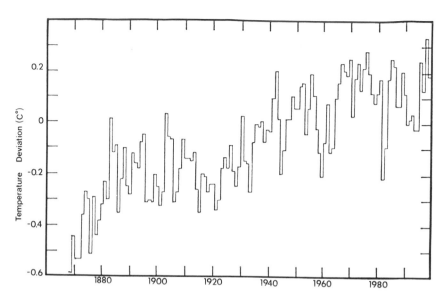

Figure 3
Changes in global mean temperatures since 1860.

near as marked as the apparently continuing decrease in mean temperature in the northern regions. A reverse of the warming trend seems to have occurred. Since this time, however, average global temperatures seem to have been increasing.

Averages, however, are often misleading. A meteorologist friend, Dr. James Lodge of the National Center for Atmospheric Research in Boulder, Colorado, continually reminds himself of this fact with a cartoon on the wall of his office. This cartoon shows a man with his head in an oven and his backside in a refrigerator and a rather dim technician measuring the temperature of his navel. The caption reads "His average temperature is fine, Doc." So it is with the world. The world's mean surface temperature may not have changed much, but large fluctuations are evident from year to year on a local scale.

Not only is the mean surface temperature in the northern hemisphere changing; other properties are also changing. One is the frequency of westerly wind drift across the North Atlantic. Hubert Lamb, of the British Meteorological Office, has compiled records of westerly wind frequencies going back to as early as records were taken in the past century. The decline of the westerlies in the early 1970s was greater than had ever been observed before. The decline in temperature at this time and the decline in westerly winds would expectedly be associated because the warm winds traversing the Atlantic toward the Arctic Basin are westerlies, and if such winds become infrequent, northern hemispheric temperatures, at least in the regions above the British Isles, become lower over time.

The extent of these climatic changes seems significantly greater than "the climatic noise level." Explanations fall into three categories. The first is "natural/organic." Some authorities suggest that the decline in northern hemispheric temperatures since World War II and through the mid 1970s is largely attributable to increased volcanic activity. Others attribute changes to solar output. Volcanic dust veils had occurred, but fewer were present then than in the nineteenth century. They seem to me insufficient to account for this episode of rapidly declining temperature. The decline was occurring considerably earlier than the most severe of the volcanic dust emissions, e.g., from Mount Agung in Bali (1963–64). The sun's output, which changes little over the sunspot cycle (approximately eleven years), is quite insufficient to account for surface changes on Earth.

Another category of explanations is anthropogenic; it includes factors such as carbon dioxide from fossil-fuel emissions, jet-aircraft contrails (which reflect sunlight back into space), and human dust-raising activities in general, including farming, slash-and-burn agriculture in the tropics, and the production of aerosols as a result of the burning of sulfur compounds. Anthropogenic changes are our own contributions to the environment, and climatic changes may be a response of the global biota to anthropogenic perturbations.

Controversy is intense around the issue of possible effects from industrial, domestic, and agricultural activity. By the early 1970s more than 15 billion tons of carbon dioxide had been injected into the atmosphere, and at least half of this increment still circulates. The rate of increase of carbon dioxide concentration has accelerated so much that it cannot be accounted for merely by emissions.

Meteorologists are in unanimous agreement that CO_2 increase leads to a warmer climate, especially in northern latitudes. Exactly the reverse could happen, however. Both dust and haze reflect sunlight back into space, reducing the heat received from the sun and causing the Earth's surface to cool. Cooling has occurred, but this simplistic explanation does not hold. Hazes in the troposphere absorb more energy than they reflect; a haze in the stratosphere may warm that region also by absorbing energy, but this results, by a strange meteorological sequence, in the cooling of the Earth's surface such as happens when volcanoes inject haze into the stratosphere. Haze near ground level should have a slight warming effect. Almost every man-made effect, from jet contrails to carbon dioxide emissions and haze generation, should cause a rise in mean surface temperature. Obstinately, though, the temperature during this time fell.

The Challenger expedition in the 1950s showed that biological systems readily add methyl radicals to a wide variety of elements, including sulfur. I thought that dimethylsulfide (DMS) might be the major biological sulfur compound emitted into the atmosphere. Challenger had already shown that marine algae and certain land plants emit dimethylsulfide. The production of dimethylsulfide is ubiquitous in the biosphere. Marine algae, soil, and almost all plants emit it. The output of DMS is strongly light-dependent; it may be the missing component of the natural sulfur cycle. It is far more stable than hydrogen sulfide, and so could survive transfer to the stratosphere, where it might oxidize to give sulfate and perhaps a methane sulfonate aerosol. Such an aerosol is well worth seeking; if found, the source will be biological.

Another product of the biota tantalizingly suggestive of important atmospheric change is nitrous oxide (N_2O). N_2O is emitted, mostly by soil microorganisms, at the huge rate of 2.5 billion tons per year. The concentration of N_2O in the troposphere is about 0.5 ppm. In the stratosphere the destruction of N_2O by solar ultraviolet light leads to the production of, among other things, nitric oxide (NO), which modifies ozone production. Nitrous oxide production may affect the density of the ozone layer; it may be another biological climate modifier (figure 4).

Regarding recent climatic change and the possible responsibility of man's activities, facts are few and opinions many. But while I think it will be a long time before the complete system that determines the climate is understood, the answers are unlikely to be found if the biota is neglected. The biota has continued to survive and modify its environment for more than 3 billion years. Changes which have occurred since the evolution of *Homo sapiens*, and especially more recently since the industrial revolution, seem huge to us; the consequences seem dire and immediate. But devastating environmental change due to rapidly growing populations of young species is a recurring theme of the

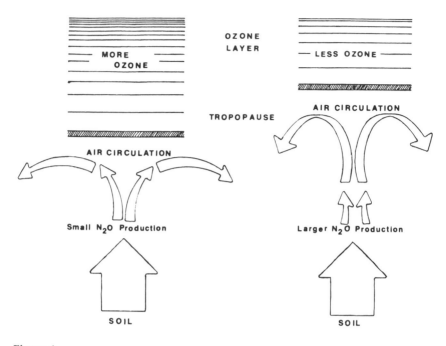

Figure 4
Nitrous oxide (N_2O), produced by soil microorganisms, can modify the ozone layer, causing changes in atmospheric circulation patterns.

evolution of life. A strong response of exponentially increasing organisms to their own "pollutants" has happened before. The enforcement of gas-emission standards on internal-combustion engines, the building of "biospheres" by people in Arizona and in the USSR, the announcements both by NASA and by Earthwatch of their plans for "missions to Planet Earth," and the great rise of environmental concerns on the part of industry and educators all begin to exercise a negative feedback on the tendency of our global human population to make its immediate environment unfit for our kind of life. This behavior is typically "gaian"; indeed, the hope I want to leave you with is that Gaia, in fact, truly exists.

Gaia: What's New?

The first presentation on the Gaia hypothesis for our Environ-mental Evolution course was made in 1973. Now everyone wants to know what's new. What is its current status?

A great deal of new evidence has accumulated. We now have a quite plausible theoretical model of the way Gaia works. A good ex-ample of this is the close coupling between carbon dioxide and cli-mate regulation. There is little doubt that increasing carbon dioxide in the atmosphere increases the absorption of outgoing infrared ra-diation through the greenhouse effect and tends to warm the planet. The evidence for this has strengthened over recent years. Exciting new evidence in the last couple of years has come mostly from analysis of ice cores from Greenland and Antarctica. A geologic re-cord of the carbon dioxide concentration of the atmosphere from the present to tens of thousands of years into the past (well into the last glaciation) can be documented from gas samples from ice taken at different depths. The exciting aspect of this new information is the correlation between CO_2 and temperature. During the last glaciation, the CO_2 concentration fell to somewhat below 200 parts per million; temperature-CO_2 correlation is very close to that of the model predic-tions made beforehand. Even more exciting: at the end of the last glaciation the carbon dioxide concentration rose close to its present value in a period just short of 100 years! Geophysical and geochemi-cal processes that presumably control atmospheric CO_2 concentra-tion cannot operate at that speed. In my view, this change in CO_2 concentration was a consequence of the growth and change in popu-lation density of the nonhuman biota.

What is the current status of your theoretical work on the Gaia hypothesis?

When the hypothesis was introduced, we felt some biological system must regulate the chemical composition and climate of the planet, but we did not know how. Most of us imagined a very complicated, intricate affair. I thought it might involve something built into the ge-netic structure of organisms. I am now happily confident that Gaian regulation is a natural and simple consequence of intrinsic proper-ties of life on this planet.

Three fundamental aspects of life determine its tendency to establish a Gaian regulating system. The first is the important fact that life on Earth is strongly constrained by its environment. Life does not flourish when water is frozen, nor when conditions are too hot. More favorable conditions exist between these extremes. Such a constraint applies to all manner of other planetary variables, like acidity; life will not flourish if conditions are either too acid or too alkaline. Life prefers neutrality. Ocean salinity is another constraint. If water is too saline, life cannot grow; if water is so fresh that nutrient salts are lacking, equally life will not continue. Between the extremes lie the best growth conditions. The most important property determining Gaia is the existence of life in the universe of constraints. The second crucial fact is the tendency of all life to grow exponentially whenever or wherever a niche is open, and whenever the environment becomes favorable. The third property relates to diversity. When different organisms emerge, they use opportunities when a new niche opens in different ways, or exploit old niches when other organisms fail to occupy it. So my theoretical approach is based on life's tendency to grow exponentially, limitations to this growth, and organismal diversity. From these assumptions my colleagues and I have been able to produce a simple model of the workings of Gaia. The numerous forms of life interact with their environment in an unbelievably intricate manner in the real world. It is quite impossible, even with the largest computers available, to adequately build a model of the entire world. However, we can investigate the situation by a process of reduction. I have reduced the environment to a single variable—temperature—and the species to a single type—a daisy plant—in order to produce an imaginary world, Daisyworld.

Imagine a planet very like the Earth in many ways, although with less ocean. It is well watered, and plants grow almost anywhere on its surface. It also has a very clear atmosphere, uncomplicated by clouds or greenhouse gases. The surface temperature of the planet is very dependent upon one property only: its albedo, i.e., reflection of sunlight back into space. This imaginary planet, Daisyworld, is at the same orbital distance from its star, identical to our own sun, as the Earth is from the sun. Daisyworld's sun shares a universal property: with age it grows warmer and its output of heat increases. I want to demonstrate how the temperature of Daisyworld varies with and without life as its sun increases its output of heat. The relationship between the growth of daisies and temperature is represented as a parabolic curve. Growth begins at a temperature of about 5°C and increases steadily to a maximum at about 20°C, room temperature. As the temperature rises beyond that, the growth rate declines until all growth ceases at a temperature of 40°C. This choice of growth curve is not arbitrary; it adequately describes growth as a function of temperature for most vegetation.

Growth of daisies has an effect on the environment of Daisyworld. A "species" of daisy that is light-colored and reflects sunlight tends to lower the planetary temperature. When a maximum number of light-colored daisies covers a large proportion of planetary area, the temperature of the planet is at its lowest. Conversely, when very few or no light-colored daisies exist, the planet is darker because of lack of the light reflection tendency, and the planetary temperature is much higher. Small temperature changes take place on Daisyworld depending on the population of daisies. The daisies have the capacity to regulate planetary temperature. We can simulate a lifeless world by holding the daisy population constant, not allowing it to vary. In this case large temperature changes occur with changes in solar output. A lifeless planet does not have the capacity to lessen changes in temperature brought about by increases in solar luminosity. The mean temperature of a lifeless planet can be compared with one inhabited by dark daisies. Assuming that Daisyworld were a lifeless planet covered with just bare rocks, mean planetary temperature would increase as solar luminosity increased. Temperatures would rise from below freezing to about 50–70°C as the solar output steadily rose. Growth of dark daisies would have quite a different effect on temperature as solar luminosity rose. As the planetary temperature reached 5°C, dark daisies would start to grow. Imagine that dark daisies start to grow; because the stand is dark, the daisies will be warmer than their environment. This results in more growth and a little faster spreading. Before long dark daisies will cover a whole area; the temperature of that area will be warmer. This warmth adds to the extra warmth of the daisies. So, with positive feedback, daisy growth rapidly explodes until dark daisies cover a sizable proportion of the planet. The planetary temperature zooms up to close to the most favored value for daisy growth (20°C). It does not continue to rise because daisy growth is discouraged when temperature rises too high, so the planetary temperature remains more or less constant over a wide range of solar luminosity. Dark daisies alone can regulate planetary temperature to a considerable extent.

We now add light daisies and see the effect of competition between two different daisy types. Light daisies prefer to grow at warmer temperatures than dark daisies. When the mean temperature of the planet reaches about 5°C daisy growth commences. The dark daisies grow like mad until the temperature has risen to just above the optimum for growth of dark daisies on the planet. The planet is now warm enough for light daisies to grow, in spite of the fact that their tendency to reflect sunlight makes the environment somewhat cooler. This early stage, favorable for dark daisies, is less favorable for the light ones. As the sun warms up further, the two daisy populations change in number. Dark daisies decrease in numbers as white daisies increase until their growth curves intersect: a point is reached where the light daisies are just as numerous as the dark ones. As the sun warms further, white daisies are more and more favored. Still, they regulate the planetary temperature, and the mean is held very close to the most favorable value for plants. Eventually the sun becomes too hot and the entire system suddenly dies. A sudden rise in planetary temperatures occurs as all life ceases. That is the end of Daisyworld.

Daisyworld is contrived and artificial; it bears little relationship to a real planet. I do believe, however, that in principle the operation of Gaia as illustrated on Daisyworld is a close parallel to what is occurring on our own planet. Indeed, some recent evidence on the possible significance of tropical forest ecosystems and the regulation of planetary and environmental temperature is like that of Daisyworld. In the early days, I wondered why tropical forests were so dark. Their very darkness would make them absorb sunlight, like the dark daisies. Since these are the hottest regions of the Earth, it seemed to me a counterproductive trait to have evolved. Quite recently, what we should have known from the beginning has become obvious. If seen from space, these tropical forest regions of the Earth are not dark at all, but blindingly white. They are covered by white clouds that are the product of evapotranspiration from the tropical forest

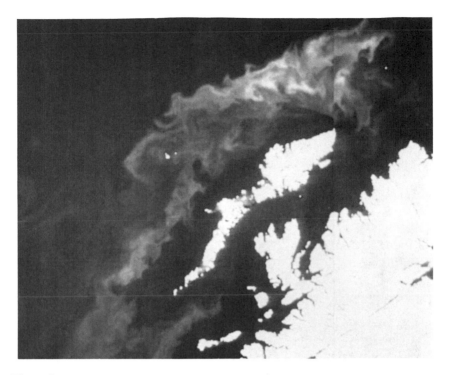

Figure 5
Patrick Holligan's photograph of a coccolithohorid bloom.

trees beneath. These clouds stabilize forest temperatures. Moreover, at night, when the sun is not shining, the clouds tend to disperse, and the very dark color is useful in dissipating the heat that is gathered during the daytime. The tropical forest—dark trees and white clouds—acts toward the regulation of the planet as both dark and white daisies simultaneously. My suggestion is speculative, and I mention it to illustrate how Daisyworld might be extrapolated to Earth. In addition to this tropical forest cloud effect, satellite photographs of oceans show blooms of coccolithophores, appearing to act in the manner of white daisies, that may affect the sea temperature (figure 5). I expect this story to develop more subplots when I come back and repeat it in 10 years.

We should consider more than the effect of albedo in the Daisyworld model. This same sort of environmental and growth feedback model helps explain regulation of concentration of atmospheric carbon dioxide. The growth of the biota continuously pumps carbon dioxide out of the atmosphere into the soil. The concentration of soil CO_2 is 30 times greater than that of the atmosphere. Similar CO_2-pumping processes occur in the sea. The biota continuously pumps CO_2 and maintains current atmospheric levels, which is probably a major climate-regulating mechanism.

In your books *Ages of Gaia* (1988), and even earlier in *Gaia: A New Look at Life on Earth* (1979), you mention the possibility of a direct connection between the Gaia hypothesis and the phenomenon of plate tectonics. Please explain.

When I prepared this material, I admit the idea was even beyond the category of speculation. Nevertheless, I stuck my neck out and included it. I am glad to report that no less a figure than Don Anderson, Professor of Geology at the California Institute of Technology, in an article in *Science* (Anderson 1984), stated quite specifically that we should consider the possibility that biological influence on the production of eclogite and limestone made plate tectonics possible. We had one closely related notion 10 years ago: that limestone can be a fluxing agent. For movement of the massive tectonic plates, the region of molten material (the magma beneath them) must be fluid and capable of motion. The nature of limestone is such that it lowers the melting point of the rock mixture into which it is drawn. This extra fluidity, present as a result of the subduction of limestone, lubricates the movements of the plates, making the process possible. The limestone—nearly all of which is biological in origin—acts as a fluxing agent, keeping the rocks below the crust molten so that heat is more readily transferred. Convection currents persist. I can't give geological details of the eclogite argument, because it is not my expertise, but I strongly advise reading Anderson's article.

Please explain the relationship between atmospheric CO_2 and limestone.

The amount of CO_2 in the atmosphere depends entirely upon its lithospheric sources and sinks. The ultimate source of CO_2 is outgassing from volcanoes, and the sink for CO_2 is calcium silicate in rocks. In a process called *weathering*, calcium silicate (a very common mineral in igneous and metamorphic rocks) reacts with atmospheric CO_2 in the presence of water and forms calcium bicarbonate and silicic acid, both of which are soluble. These are transported by rivers to the ocean, where the calcium bicarbonate dissociates. The bicarbonate is taken up mostly by organisms in the formation of calcium carbonate shells, skeletons, scales, and other biogenic structures. After death, carbonate sediments form from the rain of dead coccolithophores, foraminifera, marine animals, etc. Eventually, under pressure, limestone—primarily $CaCO_3$—forms from the former skeletal material. Later it is subducted to become the fluxing agent I just mentioned. For a better idea of the role of the biota in the formation of minerals I strongly recommend the book by Lowenstam and Wiener (1989). They give us great insight into the mechanisms of Gaia.

What new ideas have you been entertaining?

Perhaps Earth's water has been retained by Gaia. Ocean salinity, water retention, and lateral movement of crustal plates are the ideas concerning us these days. Maybe you can help us work on them!

Readings

Anderson, D. L. 1984. The earth as a planet: Paradigm and paradoxes. *Science* 223: 347–355.

Lovelock, J. 1979. *Gaia: A New Look at Life on Earth.* Oxford University Press.

Lovelock, J. 1988. *The Ages of Gaia.* Norton.

Lowenstam, H., and S. Wiener. 1989. *On Biomineralization.* Oxford University Press.

Schneider, S., and P. Boston, eds. 1991. *Scientists on Gaia.* MIT Press.

Volk, T. 1998. *Gaia's Body: Toward a Physiology of Earth.* Copernicus.

2 Comparison of Planetary Atmospheres: Mars, Venus, and Earth

Michael McElroy

To understand the history of the effects of life as a planetary phenomenon, it is important to recognize Earth as a rocky planet of the inner solar system. Were it not for life, Earth would have an atmosphere much more like those of Mars and Venus. To help us factor out the importance of the planetary background and understand the extent to which Earth is still a typical inner planet, Michael McElroy explores the salient facts about the atmospheres of our neighbors in the solar system. Dr. McElroy, Abbott Lawrence Rotch Professor of Atmospheric Sciences, is the chairman of the Earth and Planetary Sciences Department at Harvard University.

The planets of the inner solar system—Mars, Venus, and Earth (figures 1–3)—appear to share a common origin. All three were formed, in relative close proximity, from the same giant gas cloud. Present differences between these planets, then, seem to be due more to their paths of evolution than to their origin. Radioactive decay, volcanic eruption of gases, and varying levels of sunlight received and retained have shaped the compositions and conditions of these planets' atmospheres since their formation. We must recognize from the start that life makes Earth unique. Living processes exert a major influence on the composition of Earth's atmosphere and may even control its climate. As we try to understand Earth better and to predict its future course, we need an idea of what our planet would be like in the absence of life. Studies of our planetary neighbors provide a broad context for in-depth observations and analysis of the Earth.

Figure 1
Mars as seen through a telescope on Earth. The seasonal polar ice cap in the northern region of the planet is visible. *(NASA)*

Figure 2
Venus viewed from the Mariner spacecraft. The entire planet is covered by dense clouds with complex flow patterns, which obscure any surface features. *(NASA)*

Figure 3
Earth as seen from the orbiting Apollo spacecraft. It is possible to discern the continents, clouds, and an abundance of liquid water on the surface. *(NASA)*

Formation of the Planets

This story begins 4.5 billion years ago. The solar system is an embryonic cloud of hot gas and particulate materials spinning around the protosun. The planets have not yet formed, but their evolution is already underway. Because heat energy radiates into space largely from its outer boundaries, this solar nebula is not uniform in temperature. The material in the center of the cloud retains heat while the nebula becomes cooler toward its edges. As the entire nebula gradually cools, the refractory elements—those that condense at the highest temperatures and generally form heavy compounds—begin to condense. Because of the temperature gradient near the interior of the nebula, these refractory elements tend to solidify. Mercury, the densest planet in the solar system, forms during the earliest stages in nebular condensation, when conditions are ideal for the formation of dense refractory elements. In accordance with that pattern, Earth is less dense than Mercury, and Venus even less dense than Earth.

Refractory material forms the building blocks of the inner planets, condensing initially in the form of very small chunks. These blocks condense together over time. Imagining a motion picture of the solar system forming, we could see planetary formation underway at various points in the nebula. Mercury would form at the inside, perhaps near where it exists today. Venus would form a little more slowly. Further out we would see Earth, and further out still Mars. The formation of the less dense, more gaseous outer planets would just be underway. Only as the nebula cools, toward the final stages, do the most abundant elements—hydrogen, carbon, nitrogen, and oxygen—finally condense. At the temperatures we are accustomed to on the Earth's surface, these elements are not generally present in their solid form. They tend to form gases and liquids, or weakly stable solids which can be readily burned and converted to gases. We consider hydrogen, carbon, nitrogen, and oxygen to be volatiles because they tend to fly away and fill the entire atmosphere. At this stage the planets in the inner solar system are, for the most part, solid. Low-temperature condensing material, rich in hydrogen, carbon, oxygen, and nitrogen, collects around each planet and will eventually form an atmosphere and oceans. As this material rains down, it is bound up in carbon-rich meteorites called *carbonaceous chondrites,* which continue to fall on the planets today.

The second phase of planetary evolution begins as the inner planets begin to acquire more of their own character. The radioactive elements of each of the protoplanets begin to decay and to generate elements that were not present in the initial nebula. The decay of potassium forms argon; uranium and thorium decay and produce helium and other elements. A tremendous amount of heat is released in these radioactive decay processes. This vast source of energy, similar to the internal energy of the sun, profoundly changes these early planets. The planets begin to cook from the inside out and to reorganize themselves: iron and other heavy elements sink to the center to form the core, and lighter elements rise toward the surface. The lighter gases are driven into the atmosphere by volcanoes fueled by this internal energy source. By the end of this second phase, planets such as Earth evolved roughly to their present configurations.

Volatile Inventory of Earth

Having no direct information on these early days of terrestrial history, we rely on inferences based on present-day observations of Earth and other planets. We would like to have a model of evolution that encompasses all the planets. Comparative studies of Mars, Venus, and Earth help us refine our hypotheses. Starting with Earth, we note the composition of the atmosphere, the oceans, and the sediments, trying to reconstruct its volatile inventory. The most abundant volatile on Earth today, and presumably in the beginning, is water—the oceans extend over three-fourths of the Earth's surface to an average depth of almost 4 kilometers.

Perhaps not so obviously, the second most abundant volatile on Earth is carbon. Although there is relatively little carbon dioxide in the atmosphere, and the amount of carbon in living things on the surface is small in comparison with the amount of water in the ocean, we find that the major reservoir of carbon is in the sediments. There, carbon exists in the form of various carbonate minerals, including skeletons of once-living organisms. Most carbonate minerals are tied up with life histories, because they were once part of the mineralized structures of living organisms.

The third most abundant volatile on Earth is nitrogen. Here the atmosphere becomes a primary player: most of the nitrogen that was volatilized by the primitive Earth is still present in the atmosphere. Nitrogen found in living organisms is just a small component of the total nitrogen inventory, and even the sediments do not contain very much. Of the volatiles on Earth, nitrogen tends to be present in the gaseous phase, whereas carbon is present in weakly stable solids (such as carbonates) and water mostly as a liquid.

With this picture of Earth's volatile inventory, we propose the hypothesis that the volatiles on Mars and Venus began with the same relative composition as the volatiles on Earth.

Evolution of the Martian Environment

Let us consider what we would expect Mars to be like. Mars should have evolved primarily water, with carbon as the second most abundant volatile, followed by nitrogen. NASA's Viking spacecraft, which

Figure 4
The surface of Mars. The Viking lander sits in the foreground of this view of the oxidized,
iron-rich regolith (loose, rocky covering) on Mars. "Big Ben," a large rock, is seen in the
foreground. (NASA)

landed on Mars in 1976, was sent primarily to search for the presence
of life and to document conditions on the planet. Equipped with
elaborate scientific instrumentation, Viking was capable of making the
first direct measurement of Mars's atmosphere. Photographs of the
Martian surface taken by Viking's automatic cameras revealed a barren
landscape with oxidized, iron-rich rocks and sand. A sky full of gas
and particulate surface material raised by the winds scatters light
brilliantly.

Considering our hypothesis that the distributions of volatiles should
be relatively equal on the inner planets, we would have expected Mars
to have released enormous amounts of water. But on Mars that water
would not form an ocean—at least it certainly would not today. Since
Mars is further from the sun than Earth, it is too cold for water to be
present in liquid form. The water is present as ice or is bound up in
the mineral structures of the planet's upper layers. Even if primitive
volcanoes on Mars had released large amounts of liquid water, it
would simply have frozen or been incorporated into rock. We infer

that Mars does have abundant water—perhaps even more than Earth has—in the upper layers of its lithosphere.

Carbon, the second most abundant volatile element on Mars, follows a similar scenario. Some of it is present in the atmosphere as carbon dioxide. In fact, Mars has much more carbon dioxide in its atmosphere than Earth. All things being equal, this would make photosynthesis an easy process on Mars. As it does with water, Mars's temperature limits the quantity of atmospheric CO_2. The polar cap contains frozen carbon dioxide as well as frozen water. If Mars were to warm up by ten or twenty degrees, it would rapidly acquire an atmosphere containing even more carbon dioxide and would have a surface pressure comparable to Earth's.

We expect nitrogen to be present on Mars mostly in the atmosphere, as on Earth. If we assume not only that the relative abundance of water, carbon, and nitrogen is the same on Mars as on Earth but also that the total masses of these elements are comparable on the two planets, we encounter the first problem with this universal hypothesis: We expect a very dense atmosphere of nitrogen on Mars, but this is not the case. The atmosphere of Mars is composed primarily of carbon dioxide. Nitrogen, though still the second most abundant constituent of the atmosphere, represents only 3 percent of the total atmosphere relative to CO_2. How do we resolve that problem? Since Mars is less massive than Earth, its gravitational field is not quite as strong. It has trouble retaining some of its gaseous elements. Nitrogen, in fact, escapes from Mars by an interesting process recently identified by scientists. Gases tend to ionize in the upper levels of the planetary atmospheres. When they neutralize one another, they break apart and release atoms at fairly high speeds. A speed of 5 km/sec is sufficient for one of nitrogen's primary isotopes, ^{14}N, to escape from the atmosphere. The ^{14}N reaches this critical speed, but the other primary isotope, ^{15}N, generally does not. Nitrogen atoms are continually and selectively driven off into space in this manner. We can use this selectivity to test our hypothesis. Since we know the relative abundances of those isotopes on Earth (^{14}N is approximately 100 times as abundant as ^{15}N), and since we presume the ratio to have been the same on early Mars, we can predict the quantity of dispersion. If escape that releases primarily ^{14}N has occurred for 4.5 billion years, we expect to find relatively more ^{15}N on Mars than on Earth. To our satisfaction, one of the first major discoveries of the Viking missions was that atmospheric ^{15}N was almost 60 percent more prevalent than we would expect its initial configuration to be.

In summary, our knowledge of volatiles on Mars is generally consistent with the view that Mars was similar to Earth in its origin but that the two planets have very different histories with respect to nitrogen. In their histories of water and carbon dioxide, Mars and Earth are more similar; temperature is a controlling factor in regulating the composition of the current atmosphere of Mars.

Volatile Inventory of Venus

Now let us consider Venus. A historic photograph of the surface of Venus was taken in the early 1970s from the first Soviet landing vehicle. More recent images of the surface of Venus have been taken from successive Soviet Venera spacecraft (figure 5). Given that the spacecraft had to withstand an atmospheric pressure almost 100 times that on the surface of Earth, these photos represent a remarkable technological achievement. The temperature of the surface of Venus, some 750°K and hot enough to melt metal, compounded the difficulty. These pictures surprised many of us in yet another respect. We did not expect enough light to penetrate beneath the very dense cloud cover of Venus to allow photographs; it seemed comparable to diving into our ocean to a depth of a kilometer to take pictures without artificial lighting. But the Soviet photographs show features of Venus, a rocky landscape, similar to places on Earth and Mars.

What did the probes discover about the volatiles on Venus? The very high pressure on the surface of Venus is exerted largely by carbon dioxide, which constitutes 96.6 percent of Venus's atmosphere. It would be interesting to compare CO_2 levels on Venus and Earth at typical Venus temperatures. If we took all of the carbonate minerals in the sediments on Earth, cooked them, and converted their carbon into carbon dioxide, carbon dioxide would become the most abundant constituent of Earth's atmosphere as well. Though Venus has almost 100 times the surface pressure of Earth, we can account for the higher pressure of Venus and for the abundance of carbon dioxide by changing our initial suggestion: Venus had a richer suite of volatiles from the start. More carbon may have been present at the time of Venus's

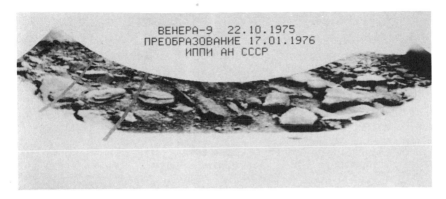

ВЕНЕРА-9   22.10.1975
ПРЕОБРАЗОВАНИЕ 17.01.1976
ИППИ АН СССР

Figure 5
Composite of photographs of the surface of Venus taken by Venera 9.

formation, or more volatiles may have been released during the second phase of planetary evolution than were on Earth. In view of Venus's high temperatures, the presence of large amounts of carbon dioxide in its atmosphere is not a surprise.

What is surprising is the absence of enormous amounts of water on Venus. On Earth and Mars, water is the most abundant volatile. We expected to find a hot, steamy environment on Venus, with a surface pressure (resulting from water vapor) perhaps 1000 times that of Earth. Yet on Venus water accounts for only about one part in 10,000 (about 100 parts per million) of the atmosphere. Where is the water? If hydrogen escaped from Venus as nitrogen did from Mars, we would expect to find Venus's current hydrogen, or its water, to be a little on the heavy side. Just as nitrogen comes in two isotopes, ^{14}N and ^{15}N, hydrogen comes in two forms: hydrogen, with atomic mass 1 (^{1}H), and deuterium, with atomic mass 2 (^{2}H). We know with confidence the abundance of those elements at formation, and we know their relative abundance on Earth by measuring their ratio in the oceans. We were delighted when mass spectroscopy revealed that Venus was in fact enriched in deuterium by about a factor of 100, which was evidence for the early escape of the lighter forms of hydrogen. Venus has the highest relative abundance of deuterium to hydrogen we know of in the solar system. Its water is rapidly switching to favor the heaviest form, rich in deuterium.

We expected Venus to have retained nitrogen; the process that allowed nitrogen to escape from Mars should not work on Venus because of the stronger gravitational field. If Venus was like Earth and Mars at time zero, its atmosphere should have a few percent nitrogen by relative abundance to carbon dioxide. Indeed it does. Nitrogen and carbon dioxide on Venus stand the test. Both support this universal hypothesis for volatiles on the planet, leaving the case of water as the only major unsolved uncertainty. Venus may or may not have formed with a volatile inventory like that of Earth. One scenario suggests that in its earliest stages of evolution Venus may have had a very hot, steamy atmosphere that was lost through evaporation to space. Hydrogen and deuterium, and perhaps oxygen as well, streamed off in an episode of planetary evolution for which we have not discovered any record. After this, another evolutionary phase began with less water, perhaps by a factor of 100, than it had at time zero. A second possibility, which I favor, is that Venus began with less water than either Mars or Earth. Water, more than the other volatiles, is most affected by changing temperatures. The proximity of Venus to the sun may have caused it to lose more of its water initially. The history of Venus is generally similar to that of Earth and Mars, but it is significantly different with regard to water.

Earth's Atmosphere as a Product of Life

As figure 6 shows, the present atmospheres of Mars and Venus are predominantly carbon dioxide whereas that of Earth is predominantly nitrogen. If Earth had not undergone an evolution that involved living organisms, its atmosphere and its surface would look profoundly different. There might be large amounts of nitrogen in the atmosphere, but there probably would be much more carbon dioxide. In contrast with the current level of less than 1 percent, carbon dioxide would be the most abundant constituent of the atmosphere. The oceans might contain most of the hydrogen released in the form of water from the interior of the planet. (The presence of oceans is more a product of the ambient temperature than of life.)

The most dramatic characteristic of Earth's atmosphere, however, is the presence of a large amount of free molecular oxygen. Molecular oxygen is the direct product of life on Earth. Photosynthetic organisms evolved the ability to build biological tissue from carbon dioxide and water using solar energy, liberating oxygen in the process. For every

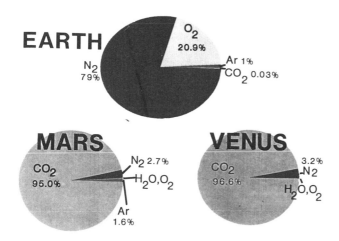

Figure 6
The major components of the present atmospheres of the terrestrial planets. The percentages represent numbers of molecules, not relative weights. Elements without percentage values are present in trace amounts. *(Jeremy Sagan, Presentation Express)*

oxygen molecule in Earth's atmosphere, a fossil organic carbon residue lies buried in sediment. Every carbon atom that is present in reduced form in the sediment has an oxygen analog in the atmosphere. The separation took place sometime in the past. When the sediment is uplifted and reoxidized, the carbon reunites with oxygen to form carbon dioxide.

Some episodes of Earth's atmospheric history are easier to make sense of than others. Predicting the future course of the evolution of our planet is difficult because of a number of major paradoxes in its history.

One such paradox is the relative constancy of Earth's temperature. Astronomers believe that at the time of the sun's formation its luminosity was substantially lower. The amount of energy the early Earth received from the sun therefore would have been much lower. Despite this inferred lower amount of solar energy, the geologic record indicates the persistence of a fairly benign global climate. A fluctuation of even 5°C in the average temperature would be capable of inducing catastrophe. The water in the oceans is poised among three phases. If warmed, the water turns to vapor (gas); if cooled, the water turns to solid ice. Water is found in its gaseous phase on Venus, and in its ice phase on Mars. Water absorbs radiation efficiently. If much of Earth's water were to evaporate, the resulting greenhouse would evaporate the remaining water more rapidly, leaving a smoldering, Venus-like surface with a temperature of 750°K—a scenario which obviously did not happen. If the oceans were to freeze, the increased albedo from the icy sheen would reflect much of the sunlight back to space, and the water would quite likely remain frozen. (J. C. G. Walker of the University of Michigan may be correct, however, when he argues that even if the oceans were to freeze over, the continuous venting of volcanic gases would produce a greenhouse effect and melt the ice.)

We know a great deal about Earth's past climates because we have the ability to look back in time in a variety of ways. A core of sediment from the floor of an ocean gives us information extending nearly a million years into the past. Sediment cores from continental areas allow us to infer conditions even more ancient. We can also draw climatic conclusions on the basis of the types of life present at various points in time; from data of this type we learn of the planet's temperature constancy despite environmental changes over time.

Mars, it seems, has undergone climatic changes—the presence of deep channels, which probably carried large amounts of liquid water, indicates that temperatures there were once significantly higher. Major climatic variation is less likely to have occurred on Venus; its dense atmosphere acts as a regulator. These observations lead to speculation as to how physical, chemical, and biological processes interact to determine and maintain a relatively constant temperature on Earth.

Comparative planetology is challenged to understand the evolutionary process that brought Earth from its origin to its current state. This challenge involves the biological aspect of the evolutionary history of the planet as well as the physical and chemical aspects. We need to continue to develop this broad perspective to better predict the changes that may occur as we continue to alter the composition of the atmosphere. The increase in carbon dioxide is but one of many global-scale human-induced changes. We mine organic residues of coal and oil, and burn them rapidly. Long-term monitoring shows that atmospheric carbon dioxide amounted to only about 280 parts per million in 1850; by the early 1980s it was 350, and it is increasing steadily. This is attributed (at least in part) to the burning of fossil fuels, and perhaps also to the deforestation of the tropics. Equally dramatic evidence exists for the change in the concentration of atmospheric methane, which is increasing by 1–2 percent per year. Large amounts of methane are produced by bacteria living in cellulose-digesting cattle and other ruminants, the population of which expands along with the human population. And our agricultural practices continue to create environments that favor methanogenic bacteria; for example, the flooding of fields in order to grow rice creates anaerobic conditions favorable to bacterial production of methane. Furthermore, fluorocarbons—for which there are no natural analogs—are appearing in the atmosphere as a result of human activity. Each of these gases can affect the climate by trapping heat radiated by the Earth's surface.

It is believed that Earth's climate has the capacity to change. If, despite the varying levels of atmospheric gases, it does not change, we would like to understand why. Thus the problems of understanding Earth's past and predicting its future merge into one grand scientific challenge of understanding the factors that control the planetary climate. Comparative planetology helps create the broad perspective needed for wise assessment of the future.

Editors' Note

Impressed by the continuing discovery of molecules in space by our University of Massachusetts colleague William Irvine of the Physics and Astronomy department, we are adding at the end of this chapter three tables kindly supplied by Professor Irvine from his paper "Extraterrestrial organic matter: A review" (Origins of Life 28 (1998): 365). These tables demonstrate that organic matter occurs in abundance around and between stars, on surfaces of interstellar dust particles, and on comets and other bodies in our solar system.

Readings

Atreya, S. K., J. B. Pollack, and M. S. Matthews, ed. 1989. *Origin and Evolution of Planetary and Satellite Atmospheres.* University of Arizona Press.

Goldsmith, D., and T. Owen. 1992. *The Search for Life in the Universe,* second edition. Benjamin Cummings.

Weiner, J. 1990. *The Next One Hundred Years: Shaping the Fate of Our Living Earth.* Bantam.

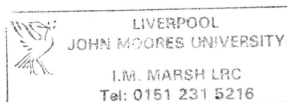

Table 1
Interstellar and circumstellar molecules.

2 atoms	3 atoms	4 atoms	5 atoms	7 atoms
H_2	C_2H	C_2H_2	C_4H	C_6H
C_2	CH_2	$l-C_3H$	C_3H_2	HC_5N
CH	HCN	$c-C_2H$	H_2CCC	CH_2CHCN
CH^+	HNC	NH_3	HCOOH	CH_3C_2H
CN	HCO	HNCO	CH_2CO	CH_3CHO
CO	HCO^+	$HOCO^+$	HC_3N	CH_3NH_2
CS	HOC^+	$HCNH^+$	CH_2CN	$c-CH_2OCH_2$
OH	N_2H^+	HNCS	NH_2CN	**8 atoms**
NH	NH_2	C_3N	CH_2NH	CH_3COOH
NO	H_2O	C_3O	CH_4	$HCOOCH_3$
NS	HCS^+	H_2CS	SiH_4^*	C_7H^*
SiC^*	H_2S	C_3S	C_4Si^*	H_2C_6
SiO	OCS	HCCN	C_5^*	CH_3C_3N
SiS	N_2O	H_3O^+	HCCNC	**9 atoms**
SiN^*	SO_2	H_2CN	HNCCC	CH_3C_4H
SO	SiC_2^*	H_2CO	H_2COH^+	CH_3OCH_3
HCl	C_2S		**6 atoms**	CH_3CH_2CN
CP^*	C_2O		C_2H4^*	CH_3CH_2OH
SO^+	C_3^*		H_2CCCC	HC_7N
$NaCl^*$	$MgNC^*$		CH_3OH	C_8H^*
$AlCl^*$	$MgCN^*$		CH_3CN	**≥10 atoms**
KCl^*	$NaNC^*$		CH_3NC	$CH_3C_4CN?$
AlF^*	HNO		CH_3SH	CH_3COCH_3
PN	H_3^+		NH_2CHO	HC_9N
CO^+			C_5H	$HC_{11}N$
SiH?			HC_3NH^+	

*: detected only in stellar envelopes
?: tentative detection

Table 2
Volatiles in interstellar grain mantles.

Molecule	Relative abundance
H_2O	100
CO	1–25
CH_3OH	<3–10
CO_2	0–15
CH_4	0–2
XCN	0–2
HCOOH	?
H_2CO	?
NH_3	?
OCS	?
H_2	?

?: tentative identification, typically at the level of a few percent.

Table 3
Volatiles identified in comets. (Does not include radicals and ions that are presumed dissociation products of volatiles present in the nucleus.)

Molecule	Relative abundance
H_2O	100
CO	2–20
CH_3OH	1–7
H_2CO	0–5
CO_2	3–5
HCN	0.05–0.2
H_2S	0.1–0.3
N_2	0.02–0.2
NH_3	trace
CH_4	trace
C_2H_6	trace
OCS	trace
C_2H_2	trace
HNC	trace
CH_3CN	trace
S_2	≤0.03
SO	trace
SO_2	trace
H_2CS	trace
HCOOH	trace
HC_3N	trace
HNCO	trace
NH_2CHO	trace
CS_2	?

3   Chemical Evolution in a Hydrogen Cyanide World

Clifford Matthews

Clifford Matthews's controversial model of an HCN world existing on the primitive Earth suggests that protein ancestors—heteropolypeptides—were formed directly from hydrogen cyanide polymers and water rather than by polymerization of individual amino acid monomers. These HCN polymers could also have served as dehydrating agents leading to the simultaneous synthesis of polypeptides and polynucleotides, precursors of today's proteins and nucleic acids. He describes the implications of this hypothesis for observations of organic matter on comets, meteorites, asteroids, and the outer planets and for speculation on the origin of life on Earth and elsewhere.

How did life begin on planet Earth? In seeking answers to this age-old question (see chapter 5), we make use of two of the most far-reaching generalizations of modern science: the unity of biochemistry and the unity of cosmochemistry.

The unity of biochemistry tells us that essential to all life are certain classes of organic compounds, in particular the macromolecular structures known as proteins, nucleic acids, fats, and sugars, familiar to each of us as daily dietary requirements. Fats and sugars—homopolymers with identical repeating units—are primarily suppliers of energy for chemical reactions, whereas the complex heteropolymer structures of proteins (enzymes) and nucleic acids (DNA and RNA) enable them to act as sources of information within living cells, the whole ensemble of these and other compounds giving rise to the characteristic activities of living systems we refer to as metabolism and reproduction. Within the enveloping framework of evolution—all living forms are interrelated by common descent—we draw the inescapable conclusion that all life must have had a common chemical origin.

The unity of cosmochemistry also arises from a kind of metabolism, this time involving the myriad stars we see out there in galaxies, as well as our parent star Sol, which brightens and, indeed, is responsible for our daily existence. It seems that a long time ago (12–15 billion years, according to current estimates) there was a Big Bang that in a matter of minutes gave rise to the fundamental particles of matter that eventually became the hydrogen (and helium) atoms that essentially constitute our expanding universe. Gravitational forces caused some of these atoms to clump together as massive conglomerates—stars— within which heavier elements soon appeared because of the intense compression. A star can thus be regarded as a factory for the stepwise production of elements from hydrogen, with the concomitant release of enormous amounts of energy we perceive as radiations of various wavelengths including heat and light. In 1967, when Hans Bethe received a Nobel Prize in Physics for first proposing these ideas, he ended his acceptance speech with the following words: "If all this is true, stars have a life cycle much like animals. They are born, they grow, they go through a definite internal development and finally die, to give back the material of which they are made so that new stars may live." The outcome of this process of stellar evolution is that interstellar atoms and molecules are widely distributed throughout the universe, giving rise not only to new stars but also to planets, possible abodes of life. As the geologist Preston Cloud put it, extending Hans Bethe's original statement: "Stars have died so that we may live."

Table 1 lists the relative numbers of atoms of different elements existing within and between the stars of our galaxy, the Milky Way. Only the most common elements are represented, expressed on a percentage scale, the total of all atoms being 100. Most striking is the fact that in this hydrogen-rich environment the next most abundant elements (other than the inert gases helium, neon, and argon) are oxygen, carbon, and nitrogen. The hydrides of these elements i.e. H_2O, NH_3, CH_4 are among the many molecules so far detected in dense molecular clouds between and around stars, many of these having been formed by photochemical reactions around interstellar grains possessing inorganic and organic components (see table 2; see also Irvine 1998 and the editors' note to chapter 2 above). These same four elements—H, O, C, and N—are also the major components of living organisms, constituting 99.5 percent of the biosphere. Why this amaz-

Table 1
Cosmic abundances of most abundant elements (percentages of atoms).

Hydrogen	87.0
Helium	12.9
Oxygen	0.025
Nitrogen	0.02
Carbon	0.01
Magnesium	0.003
Silicon	0.002
Iron	0.001
Sulfur	0.001
Others	0.038

ing coincidence? The answer came in 1953 from a revealing experiment which showed dramatically how the unity of cosmochemistry could be related to that of biochemistry.

Considering the question of Earth's origin with table 1 in mind, Harold Urey concluded that the aggregation of abundant atoms such as H, C, O, and N around the developing Sun would cause planets to form with atmospheres consisting mainly of the hydrogen-rich compounds methane (CH_4), ammonia (NH_3), and water (H_2O) as well as molecular hydrogen (H_2). On Earth, no H_2 would have settled, but the other gases present—methane, ammonia, water—would eventually have been converted to today's oxidizing mixture (O_2, N_2, and CO_2) due to the presence of life itself following prebiotic chemical reactions in the atmosphere and oceans. How did this come about? What happened to the methane and ammonia?

It was questions such as these that intrigued a new graduate student, Stanley Miller, after he heard Urey lecture on these ideas at a departmental seminar at the University of Chicago. Together they carried out the now-famous Miller-Urey experiment, in which the postulated components of Earth's primitive atmosphere were subjected to an electric discharge that brought about the ready synthesis of a number of organic compounds known to take part in today's biochemistry (figure 1). Most significant was the detection of four of the twenty α-amino acids that are the building blocks of today's proteins. Later spark studies by Miller and others showed the presence of even more of these important monomers. Cyril Ponnamperuma and his co-workers in the Laboratory of Chemical Evolution at the University of

Table 2
Interstellar and circumstellar molecules.

2 atoms	3 atoms	4 atoms	5 atoms	6 atoms	7 atoms	8 atoms
H_2	C_2H	C_2H_2	C_4H	$C_2H_4^a$	C_6H	$HCOOCH_3$
C_2	CH_2	1-C_3H	D_3H_2	H_2CCCC	HC_5N	CH_3COOH
CH	HCN	c-C_3H	H_2CCC	CH_3OH	CH_2CHCN	CH_3C_3N
CH^+	HNC	NH_3	$HCOOH$	CH_3CN	CH_3C_2H	C_7Ha
CN	HCO	$NHCO$	CH_2CO	CH_3NC	CH_3CHO	CH_3C_4H
CO	HCO^+	$HOCO^+$	HC_3N	CH_3SH	CH_3NH_2	CH_3OCH_3
CS	HOC^+	$HCNH^+$	CH_2CN	NH_2CHO		CH_3CH_2CN
OH	N_2H^+	$HNCS$	NH_2CN	HC_3HO		CH_3CH_2OH
NH	NH_2	C_3N	CH_2NH	C_5H		HC_7N
NO	H_2O	C_3O	CH_4	HC_3NH^+		C_8Ha
NS	HCS^+	H_2CS	SiH_4^a			CH_3C_4CN ?
SiC^a	H_2S	C_3S	C_4Si^a			CH_3COCH_3
SiO	OCS	$HCCN$	C_5a			HC_9N
SiS	N_2O	H_3O^+	$HCCNC$			$PAHs^b$
SiN^a	SO_2	H_2CN	$HNCCC$			$HC_{11}N$
SO	SiC_2^a	H_2CO	H_2COH^+			
HCl	C_2S					
CP^a	C_2O					
SO^+	C_3^a					
$NaCl^a$	$MgNC^a$					
$AlCl^a$	$MgCNa$					
KCl^a	$NaNC^a$					
AlF^a	HNO					
PN						
CO^+						

[a]detected only in stellar envelopes
?: tentative detection

Maryland were able to detect adenine (a key component of DNA, RNA, and ATP) as well as the other four nucleic acid bases. Adenine had previously been obtained by John Oró from aqueous reactions of hydrogen cyanide (HCN) and ammonia, which also led to the synthesis of glycine and other α-amino acids. Further extensions of this HCN chemistry by James Ferris and Leslie Orgel enabled more nitrogen-containing heterocyclic compounds to be formed. By now it seems well established from a variety of experiments (see figure 2) that many of the very compounds needed for biochemistry, components of the

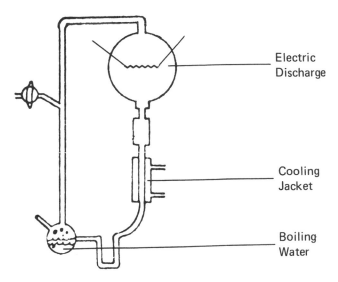

Figure 1
Miller-Urey apparatus (all glass) for subjecting a mixture of methane, ammonia, and water to a high-energy electric discharge, simulating atmospheric reactions on the primitive Earth.

essential macromolecules of life, would have been readily formed in reducing environments. How they might have got together to form the necessary polymers, however, remains problematic. Indeed, this uncertainty has led to an impasse in current thinking about chemical evolution and the origin of life, raising questions that need to be addressed.

Let us look at the situation more closely. A widely held belief is that the prebiotic formation of primitive proteins, for example, occurred in two steps: the synthesis of α-amino acids brought about by the action of natural high energy sources on the components of a reducing atmosphere, followed somehow by their linking together—condensation—to form polypeptides on the surface of planet Earth. The pioneering demonstration by Miller and Urey that α-amino acids are readily obtained from methane, ammonia, and water subjected to electric discharges, taken together with subsequent apparently successful syntheses of peptides from amino acids by dehydration reactions (by Sydney Fox and others), seems to be in accord with this view. When I look more critically at the evidence for condensation, however, I find that the specific conditions chosen—anhydrous,

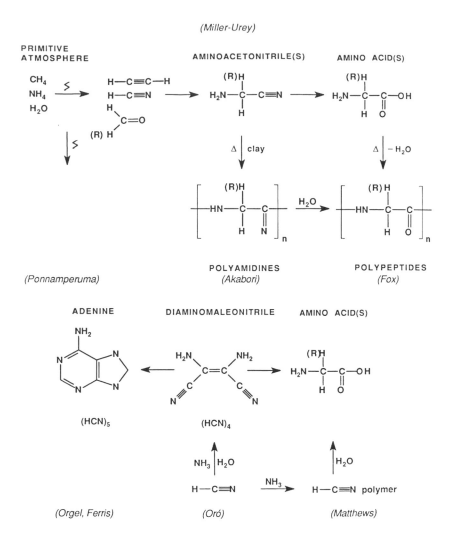

Figure 2
Prebiotic synthesis experiments, 1950–1975.

high temperature, acidic, for example—are not necessarily characteristic of a young, developing planet. Troubling questions also arise
concerning the concentration, purification, and interaction of the initial
products, a host of organic compounds constituting a dilute soup. Even
if complex α-amino acids were present, could they have joined together selectively to form long chains in amounts sufficient for life's
beginning? How plausible are these attempted simulations as models
of prebiotic chemistry? To me, and to many others, they almost suggest
that life could not have started here on Earth!

How, then, did this polymerization come about? I propose that a key
step was the synthesis of polyaminomalonitrile (see figure 3), a
polyamidine that can be formally derived by addition reactions from
aminomalononitrile, $H_2NCH(CN)_2$, itself a trimer of hydrogen cyanide.
Compare possible addition reactions of aminoacetonitrile to give
polyamidines (Akabori, figure 2). Figure 4 shows how the polyamidine
structure of polyaminomalononitrile, made up entirely of HCN molecules, can be transformed to the polyamide (or polypeptide) structure
of proteins by treatment with cold water, since as a rule, amidine
groups

—C—NH—
‖
NH

are readily converted by water to amides

—C—NH—.
‖
O

Further, the nitrile groups ($—C{\equiv}N$) in each repeating unit are so
reactive that cumulative addition of more HCN (as well as acetylene,
H_2S, etc.) can convert them to side chains, denoted here by R', to give
heteropolyamidines (that is, polyamidines with side chains that are
diverse, not merely repeated). When this modified cyanide polymer
meets water, the amidine groups become amides and R' becomes R,
where R represents the side chains of today's proteins. What is so
intriguing about the parent polymer, polyaminomalononitrile, is that
it can be converted by more HCN and, finally, water to heteropolypeptides possessing both the backbone and the side chains of proteins! In
principle, then, two of the simplest and most common molecules in the

universe, HCN and H_2O, can give rise to a variety of primitive proteins.

What kind of experiments can we perform to obtain evidence for this hypothesis? First, a variation of Miller's. We sparked a mixture of methane and ammonia (without water) and obtained a sticky brown-black solid that covered the inside of the reaction flask. Subsequent treatment with cold water yielded a yellow-brown powder. Further hydrolysis with boiling water yielded at least six amino acids commonly found in proteins—glycine mainly, alanine, aspartic acid, glutamic acid, serine, and valine—as well as some non-protein amino acids. More directly, we found that hydrogen cyanide itself (a colorless liquid boiling at 25°C), with a trace of added ammonia, becomes solid in a few hours, changing in color from yellow to orange to brown to black. Again, after extraction with cold water we obtained a yellow-brown solid that can be further hydrolyzed to give the same mixture of amino acids. A black insoluble residue is usually formed in substantial amounts at the same time. Comparable results are obtained when liquid hydrogen cyanide is allowed to polymerize in water or other solvents in the presence of a base such as ammonia or an amine. It seems probable that polypeptides are present in these cyanide products after contact with cold water, since amino acids detected by combined gas chromatography and mass spectrometry and by other techniques are seen only after breakdown by drastic hydrolysis.

A subtler approach, non-destructive analysis of the total solid product obtained from HCN, became possible with the advent of cross-polarization magic-angle spinning solid-state NMR spectroscopy. With our Monsanto collaborators led by Jake Shaefer, pioneers of this new technique, we were able to show the unambiguous presence of peptide bonds—the signature of proteins—by studying labeled polymers synthesized from equimolar amounts of $H^{13}CN$ and $HC^{15}N$ and then treated with water. Our continuing investigations of HCN polymerization suggest that the yellow-orange-brown-black products are of two main types: stable ladder structures (black) with conjugated bonds, as proposed by Völker, and polyamidines readily converted by water to polypeptides (see figures 3 and 4). A sample of polymer may possess any or all of these structures, and indeed may be regarded as a multimer rather than a polymer.

Figure 5 compares the two pathways we have considered for the origin of proteins. Which came first, amino acids or their polymers? On

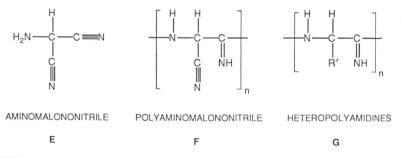

DIAMINOMALEONITRILE   POLYAMINOCYANOMETHYLENE

A   B

BLACK HCN POLYMERS (AZULMIC ACID)

C   D

HCN polymers (ladder structures) formally derived from HCN tetramer (diaminomaleonitrile) *(Völker, 1960)*

AMINOMALONONITRILE   POLYAMINOMALONONITRILE   HETEROPOLYAMIDINES

E   F   G

HCN polymers (polyamidine structures) formally derived from HCN trimer (aminomalononitrile) *(Matthews, 1966)*

Figure 3

Structures of hydrogen cyanide polymers. A sample of HCN polymer may possess any or all of these structures including hybrids (multimers).

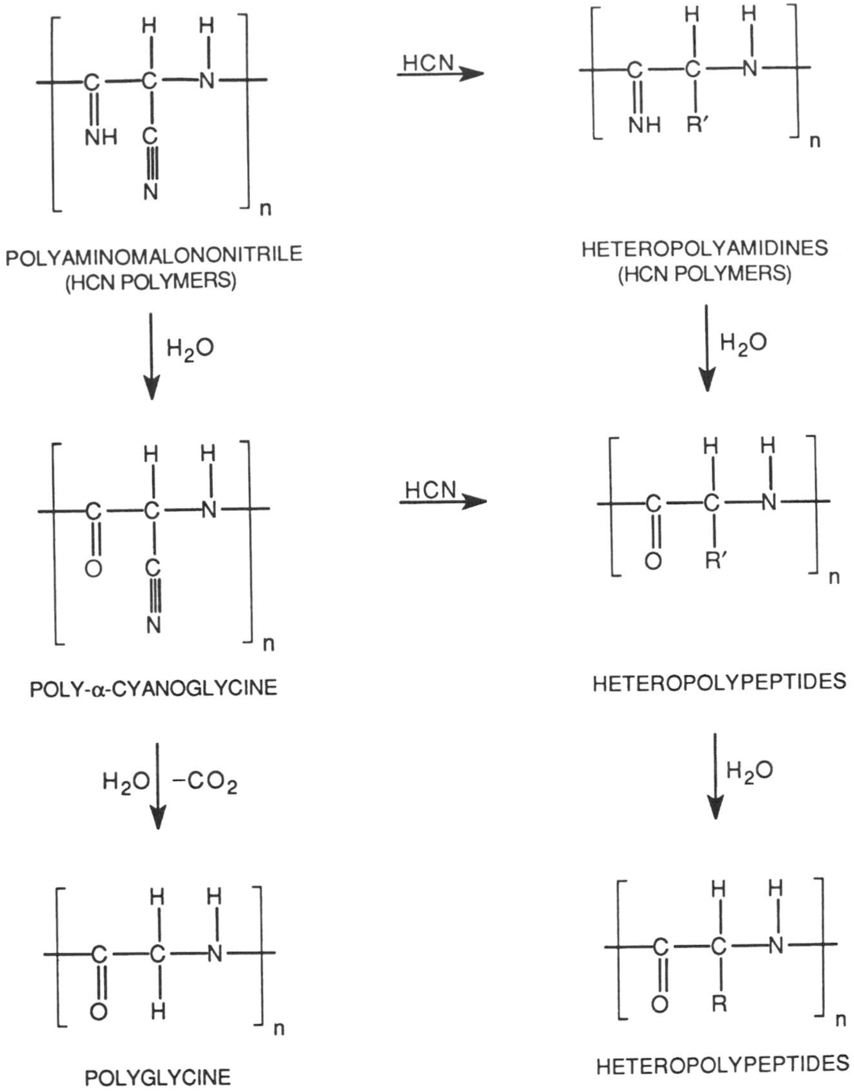

Figure 4
Polypeptides from polyamidines. Cumulative reactions of HCN on polyaminomalononitrile yield heteropolyamidines (with side chains R') which are converted stepwise by water to heteropolypeptides (with side chains R).

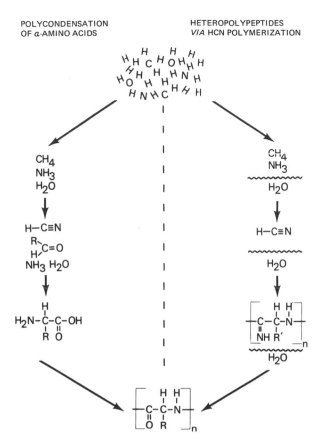

Figure 5

Two opposing models for the origin of proteins. Which came first, amino acids or their polymers? The left pathway shows α-amino acids somehow condensing in a dilute soup to form polypeptides. The pathway on the right shows clouds of HCN polymerizing in the atmosphere to form heteropolyamidines which settle in the oceans and become hydrolyzed to heteropolypeptides (primitive proteins).

the left is shown the dilute soup model whereby amino acids formed by the Miller-Urey route are assumed to have somehow got together to give primitive proteins. On the right is the cyanide model not requiring water in the atmosphere—methane and ammonia give HCN polymers directly, which settle in the oceans to become heteropolypeptides. I have argued against the first view and will now explain what I think really happened in the Miller-Urey experiment.

It seems clear to me from our reinvestigations that the primary products were not α-amino acids, as claimed, but rather HCN polymers, the HCN being formed from methane and ammonia by electric discharge reactions and by elimination from intermediates such as aminoacetonitriles and HCN oligomers (including the tetramer diaminomaleonitrile). The polymers then became hydrolyzed to amino acids, either during reflux in the reaction flask or later during the working-up procedure. The same conclusion, I believe, applies to virtually all reported experiments simulating primitive atmospheric chemistry, as well as to studies of aqueous cyanide reactions by James Ferris and others based on the original work of John Oró (see figure 2). I believe these investigations ostensibly yielding α-amino acids actually supply evidence for the abundant prebiotic existence of protein ancestors—heteropolypeptides synthesized directly from hydrogen cyanide and water.

I believe also that the cyanide model answers the questions I raised earlier arising from the dramatic results of Miller and Urey. First, why do complex molecules like amino acids appear to be more easily formed than the many simple compounds—hydrocarbons, alcohols, acids, ketones, aldehydes, amines, for example—expected from reactions of methane, ammonia, and water? Because the major intermediate, by far, is hydrogen cyanide, which then undergoes the rapid polymerization we discussed to eventually yield heteropolypeptides which can be hydrolyzed to α-amino acids. What we have here is kinetic control, with a preferred pathway defined by HCN chemistry, rather than thermodynamic control leading to a statistical distribution of products. The next question asks how α-amino acids got together to form primitive proteins. My answer, of course, is that they didn't. Instead, polypeptides appeared first, not via amino acids but from hydrogen cyanide and water. Finally, would there be enough material deposited on Earth for life to have come about? Indeed, yes, according to the cyanide hypothesis. We're not talking about a dilute soup. Instead we picture primeval Earth knee-deep in HCN polymers and, eventually, primitive proteins. Our model supplies lots of the right stuff, fast.

The ready conversion by water of polyamidines to polypeptides demonstrated by our investigations suggests that the polyamidines— HCN polymers—might have played a further essential role in chemical evolution. In the absence of water—on land—they could have been the original condensing agents of prebiotic chemistry giving rise to essential polymeric structures. Their reactive amidine groups, eager to become amides, would have brought about the stepwise formation of nucleosides, nucleotides, and polynucleotides from available sugars, phosphates, and nitrogen bases by a series of dehydration reactions. Most significant would have been the parallel synthesis of polypeptides and polynucleotides arising from the dehydrating action of polyamidines on nucleotides (see figure 6):

polyamidines + nucleotides → polypeptides + polynucleotides.

Optimum conditions might well have existed on a primitive Earth where photochemical synthesis of organic molecules proceeded in the atmosphere in three overlapping stages defined by the relative volatility of methane, ammonia, and water. As I see it, hydrocarbon chains formed first from methane by way of acetylenes. Then, as ammonia became more involved, hydrogen cyanide and cyanoacetylene became major reactants leading to the synthesis of nitriles and carboxylic acids. Polymeric peptide precursors were formed as described above together with nitrogen heterocycles possessing the basic skeletons of purines, pyrimidines, and pyridines, presumably through the ladder structures shown in figure 3. Such compounds are being obtained from HCN polymers by my collaborators Shirley Liebman and Bob Minard, using various pyrolysis techniques. When most of the ammonia had been used up, photolysis of water vapor that had been confined to lower levels became possible, leading to the third stage when oxygen-containing molecules such as formaldehyde and sugars were synthesized, as well as phosphates from phosphine. Unlike the prevailing dilute soup picture of chemical evolution, this atmospheric model supplies the necessary prebiotic compounds in sequence, in the right place at the right time. As Earth's surface became covered with this

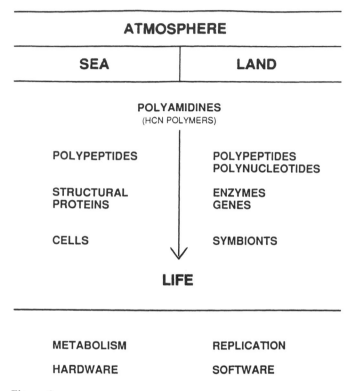

Figure 6
HCN polymers on land and sea, giving rise to life on Earth.

organic shower, potential membrane material—carboxylic acids, carbohydrates, polypeptides—accumulated in lakes and oceans, rapidly
becoming cellular, while on land the simultaneous synthesis of
polypeptides and polynucleotides, potential genetic material, was pro-
moted by cyanide polymers, perhaps assisted by clays. Interaction on
beaches of this potential "software" with the metabolic "hardware" in
protocells produced elementary replicating systems and, eventually,
the genetic code (see figure 6). On our dynamic planet this polypep-
tide-polynucleotide symbiosis mediated by polyamidines may have set
the pattern for the evolution of protein-nucleic acid systems, controlled
by enzymes, so characteristic of life today (see figure 7).

  Underlying much of past thinking about the origin of life has been
the question: Which came first, proteins (enzymes) or nucleic acids
(genes)? A popular view today is that RNA was in at the beginning,

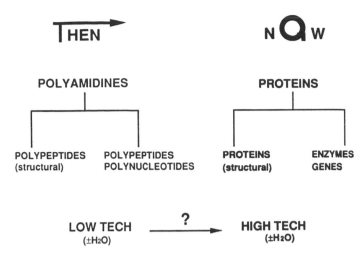

Figure 7
On our dynamic planet, this polypeptide/polynucleotide symbiosis mediated by polyamidines may have set the pattern of protein/nuleic acid systems controlled by enzymes, the mode characteristic of life today.

before DNA and proteins, since it has recently been shown that some RNA can act both as a catalyst and as a carrier of information (see chapter 5). This RNA world would eventually have evolved into today's world with its genetic code connecting nucleic acids to proteins. But how did the first RNA come about? By contrast, in the HCN world described above, HCN polymer came first, giving rise simultaneously to polypeptides and polynucleotides, precursors of proteins (enzymes) and nucleic acids (both DNA and RNA). This is an argument from abundance, given the ease of formation both of HCN and its polymers.

Let us next consider some of the extraterrestrial implications of the cyanide model. In our solar system we know that the four small, warm terrestrial planets are oxidizing in character whereas the cold, giant outer planets have reducing environments. They are the ones we're most interested in, because they still resemble primitive Earth in some important respects. For example, methane, ammonia, and water are present in Jupiter's hydrogen-rich atmosphere, as Urey predicted. What chemistry then is producing those yellow-orange-brown-red streaks shown so clearly by the Pioneer and Voyager missions? I suggest—surprise!—HCN polymerization, since these are the very colors we see in our reaction flasks. Saturn, too, could be colored with cyanide polymers, as well as Titan, its giant moon, which has an orange haze that has been shown to be polymeric. Laboratory simulations by Bishun Khare and Carl Sagan of Jupiter and Titan chemistry—reactions of methane, ammonia (or nitrogen), and water subjected to high energy sources—have yielded tholins, dark materials of undefined structure which can be hydrolyzed to α-amino acids and other compounds. We have shown by many analytical techniques (see Liebman et al. 1995) that the tholins consist largely of HCN polymers together with some hydrocarbon materials. We'll know more about this situation when various ongoing missions, in particular Galileo (for Jupiter) and Cassini-Huyghens (for Titan), send back analyses obtained by instruments parachuted into these colorful atmospheric layers.

Cyril Ponnamperuma, John Cronin, and other researchers have found amino acids and nitrogen-containing organic compounds in carbonaceous meteorites. Again, I believe these are hydrolysis (or pyrolysis) products of cyanide polymers. We have been able to extract from the Murchison meteorite the same kind of yellow-brown powders that we obtain from HCN. They yield α-amino acids following drastic hydrolysis. Similar cyanide chemistry would be expected on asteroids (the dark parent bodies of meteorites) on comets (with their frozen surfaces rich in methane, ammonia, and water). The black crust covering the nucleus of Halley's Comet very likely consists largely of cyanide polymers, a conclusion supported by the detection in its coma of free hydrogen cyanide, lots of cyanide radicals, and solid particles consisting only of H, C, and N. Most intriguing will be the analysis of cometary material brought to Earth around 2005 by the ongoing Stardust mission. I predict that HCN polymers and related compounds will be major components. It seems, then, that primitive Earth may well have been covered with HCN polymers as well as other organic compounds through bolide bombardment and/or by photochemical reactions in the atmosphere. In an aqueous reducing environment, life emerged from this vital dust, woven out of air by light.

In sum, laboratory and extraterrestrial studies suggest that hydrogen cyanide polymerization is a truly universal process that accounts not only for the past synthesis of protein ancestors on Earth but also for chemistry proceeding today elsewhere in our solar system, on planetary bodies and satellites around other stars, and in the dusty molecular clouds of spiral galaxies. This preferred pathway surely points to the existence of widespread life in the universe—protein-dominated life based on Earth-like planets possessing liquid water.

Why does the cyanide theory remain so controversial? Certainly it touches on, and challenges, much of today's research on chemical evolution. So there is a technical opposition—Show me!—which of course is very welcome since it leads to further research and additional results, pro and con. In the words of the influential philosopher of science Karl Popper: "The way in which knowledge progresses, and especially our scientific knowledge, is by unjustified anticipations, by guesses, by tentative solutions to our problems, by *conjectures*. These conjectures are controlled by criticisms; that is, by attempted *refutations*, which include severely critical tests. . . . What is important about a theory is its explanatory power, and whether it stands up to criticism and to tests."

But there is also a more profound objection, which I can best illustrate by using comments of the late Cyril Ponnamperuma. In an article in *Science News* reporting our NMR results, Cyril was quoted as finding the Matthews hypothesis "an interesting suggestion but probably one that is much too complicated." This surprises me, since if anything I would consider simplicity to be its main strength, simplicity of a type that generates complexity. Cyril went on to suggest that the logical approach is to build up more complicated structures like proteins from simple building blocks like amino acids; philosophically it seems more likely that the simple structures came before the complicated ones. Yes, but. . . . The fallacy here lies in the assumption that amino acids were the original building blocks, as they are today. I am arguing instead that the reverse is true—the buildings came before the blocks. In the beginning—at all beginnings—things were necessarily different. According to this model, then, proteins in all their diversity arose essentially from HCN polymers, a conclusion surprising in many ways, especially since today we are accustomed to thinking of hydrogen cyanide as leading to death rather than to life. But in those early times there was no oxygen-dependent life to be poisoned! Indeed, the original poison on Earth was oxygen, which can transform—oxidize—other compounds all too easily. Our metabolism today involves handling this poison through controlled, rather than random, oxidation reactions.

The following poem, *Blue Beginning*, is part 4 of a longer poem written in 1983 by George Drury, an undergraduate English major at the University of Illinois at Chicago, while taking a course for non-science majors taught by Professor Matthews entitled *Chemistry and Life*. The *Blue* refers to the Latin word *cyan,* as in *cyanide*—so named, I hear, because of the blue color brought on by asphyxiation resulting from breathing HCN. The poem is introduced by these lines from George Herbert's *Providence:* "Ev'n poysons praise thee. Should a thing be lost? Should creatures want for want of heed their due? Since where are poysons, antidotes are most: The help stands close and keeps the fear in view."

Blue Beginning
by George Drury

Hands covered with
crimson chalk dust
Professor Matthews

shakes a beaker
of bright
plastic beads

and talks
about his work.
Hydrogen Cyanide

and proteins.
A blue beginning.
A building defining

its bricks.
Not only
an incremental build

but a universe
in conversation—
searching

in all directions
in definition
of its instances.

The larger
into
the smaller.

Peace there
to be found
perhaps, but

only in
the finding.
Language needing
words.

Think of it—
from that deadly
substance,
life.

Readings

Deamer, D. W., and G. R. Fleischaker. 1994. *Origins of Life: The Central Concepts*. Jones and Bartlett.

de Duve, C.. 1995. *Vital Dust: Life as a Cosmic Imperative*. Basic Books.

Delsemme, A. 1998. *Our Cosmic Origins: From the Big Bang to the Emergence of Life and Intelligence*. Cambridge University Press.

Ferris, J. 1979. HCN did not condense to give heteropolypeptides on the primitive Earth. *Science* 203: 1135–1137.

Irvine, William M. 1998. Extraterrestrial organic matter: A review. *Origins of Life* 28: 365–383.

Liebman, S. A., R. A. Pesce-Rodriguez, and C. N. Matthews. 1995. Organic analysis of hydrogen cyanide polymers: Prebiotic and extraterrestrial chemistry. *Advances in Space Research* 15, no. 3: 71–80.

Matthews, C. N. 1992. Dark matter in the solar system: Hydrogen cyanide polymers. *Origins of Life* 21: 421–434.

Matthews, C. N. 1997. Hydrogen cyanide polymers from the impact of Comet P/Shoemaker-Levy 9 on Jupiter. *Advances in Space Research* 19, no. 7: 1087–1091.

Morowitz, H. J. 1992. *Beginnings of Cellular Life: Metabolism Recapitulates Biogenesis*. Yale University Press.

Shapiro, R. 1999. *Planetary Dreams: The Quest to Discover Life Beyond Earth*. Wiley.

4

Origins of Membrane Structure

David Deamer

David Deamer uses novel experimental approaches to the question of the origins of life. He recognizes that a critical problem is the first appearance of a material system segregated from its surroundings by a lipid membrane. In his search for the earliest membrane-bounded spheres, the ancestors of all cellular life, Deamer has extracted lipid-like substances from the only available source of concentrated extraterrestrial organic matter: carbonaceous chondrites. This material spontaneously forms membranes that can incorporate photochemically reactive pigment systems. Deamer conducts his experiments in the Department of Chemistry and Biochemistry at the University of California at Santa Cruz.

The molecular structure and biochemical functions of living cells are an important starting point for addressing questions related to the origin of life. By definition, all cells are bounded by membranes that encapsulate the macromolecular machinery of the life process. Our laboratory has been investigating the chemical and physical nature of membranes, and particularly molecules called lipids that are essential components of all contemporary cell membranes. To better understand how cellular life began on the Earth, we can first ask how membrane boundary structures self-assembled from the organic components that were likely to have been available in the prebiotic environment.

Phospholipids and Contemporary Membranes

Hydrocarbons are organic compounds that contain the elements hydrogen and carbon. The lipids of all cell membranes contain hydrocarbon chains, which constitute an oil-like layer that forms the boundary between the internal cytoplasmic compartment and the external environment in which cells live. Proteins, which provide the enzymatic and transport activities that are primary functions of membranes, are embedded in this fluid barrier. Therefore, to understand the origin of membrane structure we need to know how lipids and their hydrocarbon moieties provide the essential barrier properties of membranes. We can use a specific membrane lipid to illustrate this point.

Membrane lipids are typically phospholipid molecules, and a common constituent of most membranes is a lipid called phosphatidylcholine. The term *phosphatidyl* is derived from phosphate, one of the component groups of all phospholipids, and signifies that the phosphate is chemically bound to glycerol and choline. The glycerol is linked through ester bonds to two fatty acids, each consisting of an acidic carboxyl group attached to a long hydrocarbon chain. Other common membrane phospholipids include phosphatidylethanolamine and phosphatidylserine.

What kinds of molecules on the early Earth might have had lipid-like properties? Because they are relatively complex, phospholipids probably did not form the earliest membranes, so we can ask whether molecules simpler than phospholipids are able to assemble into membranes. The physical characteristics of lipids arise from a balance of polar and nonpolar properties. Certain molecular structures are polar, with relatively strong electrical charges expressed by their component atoms, while other molecular structures are nonpolar. Polar structures tend to be soluble in water and are usually referred to as *hydrophilic.* Nonpolar structures are *hydrophobic;* that is, they tend to be soluble in oil but not in water. Certain organic compounds such as lipids have both hydrophilic and hydrophobic residues on the same molecule and are therefore referred to as *amphiphilic.*

We know that a variety of hydrocarbons—oily molecules—are present in carbonaceous meteorites, and we can assume that similar compounds were present on the early Earth. Hydrocarbons are nonpolar molecules. However, if oxygen is added to a long hydrocarbon chain, the molecule becomes amphiphilic, since oxygen compounds typically have polar properties. All lipids have oxygen in their molecular structure, usually in the form of hydroxyl, carboxyl, and phosphate groups that are chemically linked to nonpolar hydrocarbon chains. So, to answer the question posed above, it seems likely that a variety of hydrocarbon derivatives containing polar oxygen groups were available on the prebiotic Earth and represented the first lipid-like molecules.

Self-Assembly of Lipids into Bilayers

Amphiphilic molecules have a remarkable property: they self-assemble into stable bilayer structures. When a lipid such as phosphatidylcholine is dried, for example, the lipid molecules form lamellar structures. (*Lamellar* means that the structure has layers of molecules.) If water is then added, water molecules penetrate between the lipid layers along hydrophilic planes, causing the lipid to swell. The swelling produces a variety of fairly stable structures, such as multilamellar lipid cylinders and spherules, each of which contains thousands of concentric lipid bilayers.

   Why do the lipid components form a stable bilayer structure in an aqueous environment? For thermodynamic reasons, hydrocarbon chains cannot dissolve in water, giving rise to the well-known observation that water and oil don't mix. When lipid molecules are placed in water, their hydrocarbon chains tend to stay in contact with one another, away from the water phase. This tendency, called the *hydrophobic effect,* stabilizes the bilayer structure. If lipids remain in contact with water, stable spherical structures called *liposomes* are produced by self-assembly (figure 1). Liposomes represent a good model for the first types of membranes to appear in the origin-of-life saga.

Sources of Organic Molecules on the Early Earth

The more general question of how *any* organic molecules appeared on the prebiotic Earth can now be considered. Virtually all present-day organic molecules are products of biological—including human—activity. Before the origin of life, how did organic molecules become available to start up the first biological systems?

   One source would have been chemical reactions within Earth's atmosphere and hydrosphere. This view is supported by experiments described by Clifford Matthews. A second probable source was the delivery of substantial amounts of organic material to the Earth's

Figure 1
Liposomes. Cylinders of rehydrated phospholipids eventually break to form rounded liposomes, as seen in the lower half of the photograph.

surface late in the accretion of the planet, about 4 billion years ago. Organic matter, present throughout the galaxy, forms layers on the surfaces of interstellar dust particles. The dust particles are concentrated in the molecular clouds from which the solar systems condense. During condensation, dust particles bring the organic material to the nascent solar system. Most of the dust and gas is incorporated into the sun, some into planets, and the rest into comets and asteroids.

The process of planet formation is incredibly violent. The temperatures are so high that all organic material is pyrolyzed and gases such as carbon dioxide are produced. However, the last stage of Earth's accretion probably involved a less violent infall of materials containing organic compounds: dust, comets, and asteroidal debris, all of which are known to contain significant amounts of organic compounds. During its late accretion stage, Earth had cooled sufficiently to permit some of the accreting organic material to survive at the surface and in the oceans.

It is possible to estimate the amount of organic material that might have accumulated, assuming such infall occurred. The result is equivalent to a layer approximately 10 centimeters deep over the entire surface of the Earth. Of course, this did not fall all at once, but instead over a period of several hundred million years. The annual accumulation may have been only a few molecules thick. Nonetheless, the total organic material delivered as infall would be several hundred times the amount of organic matter now present in the biosphere. From such calculations, it is plausible that a major source of organic matter on the prebiotic Earth was cometary and meteoritic infall.

Organic Structures in Meteorites

Because they represent the only available sample of the organic compounds present early in Earth's history, we are studying the properties of the organic substances in carbonaceous meteorites. Most of the meteorites that strike Earth today are from the asteroid belt. It is generally agreed that asteroids are the parent bodies of the meteorites, and that in the distant past the parent bodies underwent collisions sufficiently energetic to send fragments into Earth-crossing orbits. Actual meteorite falls have been witnessed every twenty years or so, when asteroid fragments captured by Earth's gravitational field produce a spectacular fireworks display as they enter the atmosphere and fall to the ground. The survival of meteoritic organic compounds during atmospheric entry is strong evidence favoring a meteoritic contribution to the organic inventory of the prebiotic Earth. However, because of their relative abundance, this would have been mostly in the form of microscopic dust particles rather than larger comets and meteorites.

In 1969, the appearance of a fireball in the skies above eastern Australia was followed by a thunderous explosion. More than 100 kilograms of meteoritic material fell near the town of Murchison, just north of Melbourne. The particular specimen of the Murchison meteorite with which we have worked weighs about 90 grams, and was a gift to our laboratory from the Field Museum in Chicago.

Carbonaceous meteorites are stony, rather than metallic, and several percent of their mass is present as organic carbon. The organic material is mostly a polymeric substance that is difficult to extract. Composed of aromatic molecules linked by ether and carbon-carbon bonds, it resembles refractory soil organics called kerogens. A smaller fraction is composed of simpler molecules that can dissolve in water or organic solvents.

To study the material, it was necessary to break open the meteorite. The meteorite is quite friable and crumbled easily. Small white pebbles approximately a millimeter in diameter can be seen in the freshly fractured surface. These are called chondrules and are characteristic of this type of carbonaceous meteorite. The fractured surface revealed its organic content when examined by fluorescence microscopy. Small fluorescent particles can be seen, ranging from 1 to 10 micrometers in diameter. Their fluorescence is produced by the polycyclic aromatic hydrocarbons that are present in the organic matter. The simplest (monocyclic) aromatic hydrocarbon is the benzene molecule: six carbons joined in a ring with six hydrogens attached. Two such rings joined together form a compound called napthalene; three form anthracene; and four form pyrene and fluoranthene. Polycyclic aromatic hydrocarbons such as anthracene and pyrene are highly fluorescent. That is, they glow (fluoresce) under ultraviolet light by emitting visible blue-green light. All these compounds are present in the Murchison meteorite, both free and attached to other molecules, and their fluorescence provides a useful marker for experimental analysis.

The next step was to partially purify the organic material and examine its chemical and physical properties. We used chromatography to separate the individual components according to the physical differences between the various molecules. (When a mixture is placed in a layer of chromatographic medium and a solvent is drawn into the medium through capillary action, the components usually interact to different degrees with the medium and the solvent. As the solvent moves up through the medium, the most soluble and least interactive components move farthest; the least soluble and most interactive stay nearer the origin, where the mixture was originally spotted.) Organic matter from the meteorite was extracted with lipid solvents such as chloroform, and a sample of the extract was placed at the bottom of a chromatographic plate. The chromatographic separation was carried out first in a solution of hexane and ether; then the plate was turned 90° before separation in chloroform. This technique, called *two-dimensional chromatography*, increases the resolution of the separation process.

Each of the chromatographically separated components was tested for the presence of amphiphilic molecules that might self-assemble to form membrane structures. Most of the extracted components were fluorescent, and this enabled us to visualize them easily on the chromatographic plate. Some of the separated material contained a complex mixture of amphiphilic compounds. When this material was dried onto a glass microscope slide and an aqueous solution was added, the dried material first broke up into droplets (figure 2). The droplets formed thick-walled fluorescent vesicles (figure 3). At higher magnification, much finer membranous structures were observed at the surface of the droplets (figure 4).

Electron-microscopic examination revealed that even the smallest droplets were surrounded by membranes (figure 5). The inset in the upper right corner of figure 5 is a high-magnification view of part of such a membrane showing a three-layer (dark-light-dark) structure, which is what we expected to see if a bilayer of amphiphilic molecules had self-assembled from the meteoritic components. The presence of a bilayer was confirmed by freeze-fracture electron microscopy (figure 6). (In this technique, a sample of the material is first frozen and then broken open under vacuum. A platinum-carbon replica of the fracture face is then made and viewed in the electron microscope. Because the original fracture plane tends to follow any bilayer structures that are present, the presence of fracture planes is diagnostic of bilayer structure.)

From these results, we inferred that membrane structures can form by self-assembly of relatively simple amphiphilic compounds, and were likely to be present on the early Earth. Although such materials clearly could have been derived from carbonaceous meteorites, other sources are also possible. Our results simply indicate that self-assembly of membranes is feasible, even under prebiotic conditions.

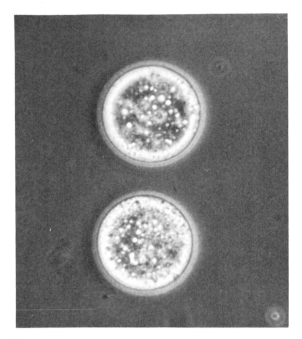

Figure 2
Chromatographically separated amphiphilic compound from the Murchison meteorite.
Rehydrated material forms droplets, as seen by phase-contrast microscopy.

Membranes and the Origin of Life

If membranes like those we make experimentally were present on the
Archean Earth, how could they have contributed to the origin of life?
The first cells required some mechanism by which a membrane could
encapsulate a system of replicating macromolecules. Although this
seems difficult, a commonly occurring drying-wetting process offers a
straightforward solution. If membrane-forming lipids are dried in the
presence of large molecules, the molecules are "sandwiched" between
alternating lipid bilayers. Upon rehydration, a substantial fraction of
the molecules are encapsulated within vesicular membrane structures.

We demonstrated this in a model system containing phosphatidylcholine and a macromolecule. We used the protein hemoglobin for the
experiment shown here, but virtually any macromolecule can be encapsulated by this procedure. The lipid-protein mixture was first dried
on a microscope slide, and a clear aqueous buffer solution was then

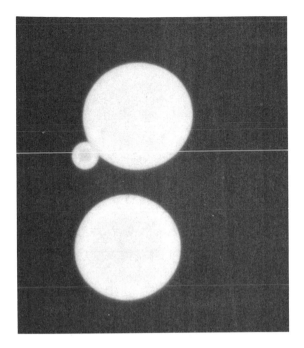

Figure 3
Droplets fluoresce under the microscope when illuminated with ultraviolet light.

added. Within a few minutes large numbers of lipid vesicles containing the protein could be observed. The colorless protein could not be seen inside the vesicles, but if we used DNA (which can be stained with the fluorescent dye acridine orange) instead of the protein, the membrane-encapsulated macromolecules were clearly visible (figure 7).

In our view, this simple one-step encapsulation is a plausible mechanism by which the first protocellular structures could have formed, probably in tidal pools subject to periodic drying and wetting. The drying process has the additional advantage of concentrating molecules from dilute solutions.

We can now return to our original questions concerning the origin of membranes and their role in protocellular structures. We have demonstrated that amphiphilic compounds in meteoritic organic material self-assemble into membranes. The chemistry of the membrane-forming components is still unknown, because even the chromatographically separated material contains hundreds of closely related compounds. However, the compounds are certainly simpler

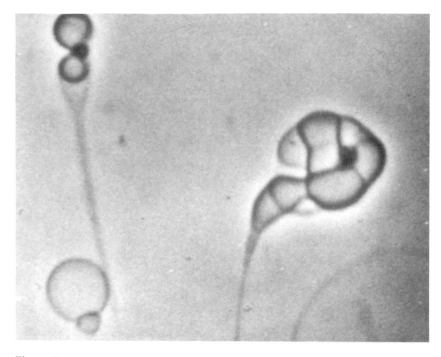

Figure 4
High magnification of material seen forming droplets in figures 2 and 3. Fine membranous structures subdivide and protrude from surface of droplets.

than contemporary membrane phospholipids, and analyses by infrared and mass spectrometry suggest a mixture of organic acids containing cyclic hydrocarbons and carboxyl groups. In earlier work we showed that organic acids as simple as decanoic acid (a fatty acid) can form bilayer membranes under conditions like these. The meteoritic amphiphiles probably form membranes by a similar process of self-assembly.

Physical processes predating the origin of life involved the accumulation of organic compounds from a variety of sources, including cometary and meteoritic infall. Chemical evolution of simpler compounds into more complex molecules, and then self-assembly of the molecules into larger structures, followed. Mechanisms to capture energy and nutrients from the environment and make them available for the growth of protocells must have been present as well.

One can imagine a vast number of natural experiments going on in the rich mixture of organic chemicals and physical environments avail-

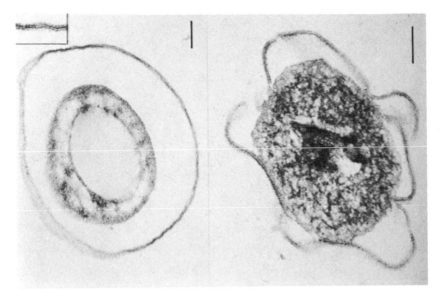

Figure 5
Electron micrographs of membranous boundary formed by lipid-like material from Murchison meteorite. "Membranes" surround amorphous material inside these drops. Scale bars = 0.1 μm. Inset shows bilayer structure of the membranous boundary.

able on the Archean Earth. We assume that at some point a molecular system self-assembled in which catalytic polymers interacted with and aided the assembly of a second class of polymers having the capacity to store information in a sequence of monomers. That sequence determined the sequence of monomers in the catalyst, producing a catalytic-information cycle. In contemporary cells, this cycle involves enzymatic proteins and nucleic acids. However, it seems likely that a much simpler cycle using RNA as both catalyst and genetic material may have been incorporated in the first forms of life.

A primary function of amphiphilic compounds is to provide closed microenvironments. If a macromolecular catalytic-information system were encapsulated within a vesicular membrane, the components of the system would share the same microenvironment. This would be a step toward true cellular function. Encapsulation would also permit a kind of speciation, in which individual cells would differ from all other cells. In this way genetically different individuals could undergo Darwinian selection based on the ability of a given cell to grow and reproduce efficiently.

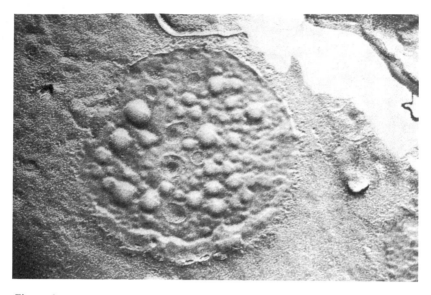

Figure 6
Freeze-fracture electron micrograph of extracted lipid-like material. The circular area represents a cast of the fracture plane indicating the presence of two layers.

A second role of early membranes was probably related to energy production, because energy-yielding processes are necessary for growth of the catalytic-information system. In contemporary cells, membranes are central to energy production. For instance, chloroplast membranes capture light energy by means of embedded pigment systems; there must have been some membrane-related process by which light energy was made available to early cells. Light energy may also have been used to drive ion transport across membrane barriers. A closed, membrane-bounded structure containing a relatively simple pigment system can provide energy in the form of ion gradients. These ion gradients could have driven selective transport processes, and later the synthetic reactions related to the chemiosmotic synthesis of ATP.

The membrane-bounded system described above encompasses a minimal set of basic cellular properties: energy and nutrient capture from the local environment, growth through a catalyzed polymerization mechanism, and replication of an information-storage molecule, all encapsulated within a membranous boundary structure. Such a system, a protocellular stage in the evolutionary process leading to the first life forms, might someday be reproduced under laboratory condi-

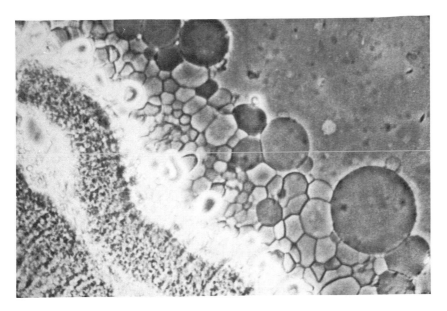

Figure 7
Phase-contrast micrograph of phosphatidylcholine-protein mixture 1 minute after hydration. Protein-containing vesicles can be seen coming from the previously dried mass.

tions. However, three central questions remain to be answered. How did the genetic code originate? Some early mechanism for coding between a nucleotide sequence and a peptide sequence must have evolved. Second, the synthesis of peptides from activated amino acids must have been catalyzed in some fashion. In contemporary cells the catalysis is carried out by ribosomes, and the origin of such a complex protein synthesis system is a major unanswered question. Last, all life today uses certain forms of amino acids and sugars called *stereoisomers* (L-amino acids and D-sugars). This appears to be a clue to the origin of life, but we still have not penetrated its significance.

Although we have no answers to these fundamental questions, the earliest prokaryotes had all the answers more than 3.5 billion years ago. The most exciting problems related to the origin of life remain to be worked out through scientific inquiry.

Readings

Chyba, C.F. and G.D. McDonald. 1995. The origin of life in the solar system. *Annual Review of Earth and Planetary Science.* 23: 215–249.

Deamer, D. W. 1997. The first living systems: A bioenergetic perspective. *Microbiology and Molecular Biology Reviews* 61: 239–261.

Horgan, J. 1991. In the beginning. . . . *Scientific American* 264: 117–125.

Kerridge, J. F., and M. F. Matthews, eds. 1988. *Meteorites and the Early Solar System*. University of Arizona Press.

5

Origins of Life: History of Ideas

Antonio Lazcano

We are almost entirely ignorant of the transition from nonlife to life. Before we knew the chemistry and the cellular structure of all life on Earth, it was easier to believe in frequent and spontaneous generation. In this chapter Antonio Lazcano presents the history of our changing concepts of the origins of life—including experimental approaches—by tracing the development of these concepts in Western Europe, Russia, and North America. He suggests that chemical and biochemical methods will solve the puzzle of the transition from nonlife to life, at least partly, by producing in the laboratory a material system that has the properties of life: information storage, replication, self-maintenance, and boundedness. Professor Lazcano teaches at the Universidad Autónoma de México in Mexico City.

Spontaneous Generation Disproved

In 1859 Charles Darwin published the first edition of *The Origin of Species*, a landmark not only in the field of biology but in Western intellectual history as well. With respect to the question of the origin of life, Darwin seemed to hold the view that the first organisms arose early in the history of the primitive Earth by a process of spontaneous generation. Darwin did not use these specific terms, but he seems to have implicitly assumed that the first living beings emerged spontaneously.

In France, also during 1859, Louis Pasteur began the experiments from which he eventually concluded that no spontaneous generation of living organisms ever occurred. In a very simple experiment, Pasteur boiled a concoction of organic materials in a flask and then sealed it. The process of boiling killed all the microbes, leaving the liquid sterile. By breaking the neck of one flask, Pasteur gave airborne bacteria access to the mixture of organic compounds, and they grew; the second flask is still sterile today, whereas the organic matter in the first was quickly decomposed. From these results Pasteur realized that bacteria did not arise spontaneously from decaying organic matter.

Pasteur's experimental demonstration of the nonexistence of spontaneous generation created a problem for biologists studying the origins of life. For example, the famous German evolutionist Ernst Haeckel had theorized that the origins of all living beings could be traced to a common ancestor; after Pasteur's compelling demonstration, Haeckel was unable to conceptualize the root of his "tree of life." Given these circumstances, many people viewed the theme of the origins of life as unscientific and therefore unworthy of study. Others, however, thought that meaningful questions could be asked and answered.

Panspermia

In 1879 Hermann von Helmholtz proposed that life had not emerged on Earth but rather had come to this planet from another part of the universe, perhaps on meteorites or comets. Helmholtz's idea was named *panspermia*, meaning "universality of the germs of life." This idea was quite popular even in the early twentieth century. More recently, the well-respected scientists Fred Hoyle and Francis Crick have tried to revive the theory of panspermia. Despite their efforts, panspermia seems scientifically moribund. The fundamental flaws in the hypothesis of panspermia are that it can never definitively be shown to be right or wrong and that there is as yet no experimental way of showing that life has indeed appeared in other parts of the universe. Panspermia does not solve the problem of the origin of life; it merely transfers the problem to the vastness of space.

Autotrophic Origins

In 1914 a group of scientists, including the physicist Leonard Troland, hypothesized a spontaneous and abiological formation of catalytic molecules in the primitive oceans. These enzymatic molecules were thought to be both heterocatalytic (capable of catalyzing reactions in solution) and autocatalytic (capable of self-catalyzation, and possibly self-replication). These ideas were further developed by the American

Figure 1
Herman J. Muller (1890–1967).

geneticist Herman Muller, who concluded that the first life form on Earth must have been a living gene able to mutate and thus to evolve. Muller's "living genes" gave rise, in his view, to photosynthetic micro-organisms ancestral to contemporary plants and, with loss of photosynthesis, to animals themselves. However, most scientists agree that no single molecule or gene can ever be "alive" outside a cell. Muller's viewpoint does not take this into account, and hence many people working in the field of the origin of life feel that it is essential to also explain the emergence of membranes and metabolism.

While Muller and Troland were developing their ideas, the Mexican scientist Alfonso Herrera proposed a theory known as *plasmogenesis*. Herrera devoted more than fifty years to experimenting with different kinds of substances, attempting to produce living photosynthetic cells in the laboratory. At first he used water and oil (or gasoline); later he experimented with hydrogen cyanide and formaldehyde. Though isolated in a conservative society that was incapable of understanding his work or even his motivations, Herrera worked tirelessly, publishing the *Bulletin de Laboratoire de Plasmogenie* and maintaining correspondence with scientists all around the world. Many of Herrera's colleagues were absolutely convinced that life could be created in the laboratory, and that the first forms of life had been *autotrophic*—that is, capable of synthesizing organic material for their sustenance.

Figure 2
Alfonso L. Herrera (1868–1942).

Formation of Organic Molecules without Life

In November 1923, a book expounding views quite different from
Herrera's was published by the young Russian biochemist Alexander
Ivanovich Oparin. Oparin's central thesis was that the first organisms
to emerge must have been bacteria in an anaerobic environment.
Contrary to most of his colleagues at the time, Oparin believed that the
first forms of life were heterotrophic—that they could not make their
own food but obtained organic material already present on the primitive Earth.

I had the opportunity to ask Oparin how his ideas had originated.
As a young student at the Imperial University of Moscow, besides
studying biochemistry, botany, and zoology he attended lectures given
privately by Kliment Arkadievich Timiryazev, a scientist who had
gone to England in order to collect as much material on Darwin and
Darwinism as he could bring back to Russia and who had later been
expelled from the University of Moscow for his political activities.
Despite his stern appearance, Timiryazev was a congenial man who
gave lectures on Darwin in his flat in Moscow. Partly because of his
influence, Oparin concluded that on evolutionary grounds it was impossible to imagine the emergence of a cell that was already fully

Figure 3
Alexander Ivanovich Oparin (1894–1980).

autotrophic. Rather, he concluded, the first organisms must have been heterotrophic.

In order to buttress his intuition, Oparin needed to demonstrate that organic material could form in the absence of living beings. Two important pieces of information supported Oparin's claim that the first organisms were more likely to have been heterotrophic. First, hydrocarbons and other organic materials were present in meteorites, and simple organic molecules in comets and in the spectra of stars. These facts had been known since the middle of the nineteenth century. Second, as the Russian chemist Mendeleev had demonstrated, long-chain hydrocarbons (oils, fats, and waxes) could be formed abiologically. On May 3, 1922, Oparin gave a lecture at the Botanical Society of Moscow in which he made his ideas on the origin of life public. His speculations were not well received.

Most of our current ideas on the origin of life can be traced to Oparin's monograph *The Origins of Life*, published in 1924 in Russian. (As a point of historical interest it is worth noting that the book, though addressed to scientists interested in the origins of life, was published primarily for the ideological purpose of spreading materialistic ideas among workers.) The Soviet revolution had taken place 7 years before the work was published; indeed, the first line of the first page reads

Figure 4
Kliment Arkadievich Timiryazev (1843–1920).

"workers of all the world unite," the famous closing line of the *Communist Manifesto*. Oparin's book remained largely unknown outside the Soviet Union, though today the seminal nature of Oparin's theses is universally recognized. Oparin proposed that there was no life on Earth when our planet coalesced. The so-called primitive atmosphere of the early Earth was reducing: hydrogen was free in the atmosphere, oxygen was present combined with hydrogen in water vapor, there was no free O_2, carbon existed in the compound methane, and nitrogen was present in the form of ammonia. In this atmosphere, ultraviolet radiation coming from the sun, electrical discharges due to lightning, and heat from volcanos all contributed to the formation of organic compounds, which later collected in the early seas to form the so-called prebiotic soup. According to Oparin these organic materials concentrated into drops, and from these drops the first cells emerged. The first cells were of course prokaryotic (bacterial) and anaerobic, since they had arisen in an environment with no free oxygen. They were probably heterotrophs that used the abiologically produced organic matter that surrounded them as a source of energy and carbon.

Contemporaries of Oparin thought along similar lines. John B. S. Haldane, a versatile British biologist, argued in 1929 that the origin of life could be explained in a context of chemical evolution. The main difference between Oparin's and Haldane's views was that whereas for Oparin the first living beings were cells, for Haldane the first living beings were viruses. Since viruses are not capable of actively maintaining themselves unless they are inside a cell, Haldane's original thoughts ran into a blind alley—the origin of the cell was not explained.

In 1936 Oparin published an enlarged version of his 1924 book. The 1936 edition was far more mature and profound in its philosophical and evolutionary analyses. Oparin not only abandoned his naive and crude materialism, he also provided a thorough presentation and analysis of the literature on the abiotic synthesis of organic material. In addition, Oparin argued that the best model of a system for the concentration of organic material on the primitive Earth, later evolving into the cell, was the coacervate (figure 5). Coacervates are charged, microscopic, organic, colloidal droplets that can concentrate organic materials existing in the medium. Since coacervates form spontaneously when two solutions of macromolecules with opposite charges are mixed, it is quite possible that they were present in the prebiotic milieu. However, they lack the lipid bilayers, present in all known cells, that retain organic matter in high concentrations inside a self-constructed boundary; therefore, they are no longer considered as potentially ancestral to life itself. Coacervates were the favorite model for a considerable time after Oparin's views became widely known, because they were perceived as mimicking the surmised properties of precellular systems. Eventually, however, it was recognized that coacervates tend to fall apart, and that their resemblance to cells is merely superficial. Experimental studies and theoretical considerations about colloids and coacervates led to their dismissal as constituting any steps toward the origins of life.

Laboratory Synthesis

For the next 10 years, while World War II raged on, nothing much happened in the field of the origins of life. After the war, Harold C. Urey returned to the University of Chicago and, having read Oparin's book, decided to analyze the primitive atmosphere from a physical-chemical point of view. Urey published his results in 1952. He suggested that the carbon, nitrogen, oxygen, and sulfur on the primitive

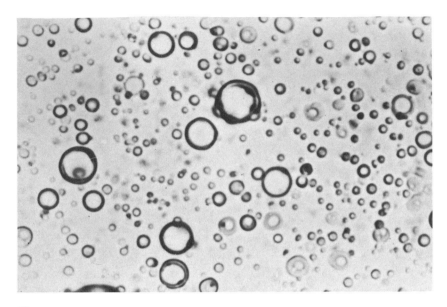

Figure 5
Light micrograph of coacervates.

Earth were probably in the most reduced forms: carbon as methane, nitrogen as ammonium, oxygen as water, and sulfur as hydrogen sulfide. Urey considered carbon dioxide (CO_2) to have been a minor constituent of the primitive prebiotic atmosphere. Stanley L. Miller, one of Urey's brightest students, designed an experiment using a "simulated primitive atmosphere" based on Urey's conception of the composition of the early Earth's atmosphere.

The question was whether prebiotic synthesis of organic molecules could have taken place in such an environment. The apparatus in which Miller and Urey simulated the primitive atmosphere was rather small and did not contain all the chemical compounds that existed in the prebiotic atmosphere. However, the experiment was ingenious—it is considered a classic to this day. The apparatus was composed of two connected flasks. Water was boiled in the lower flask; the water vapor then passed into the upper flask, where it mixed with ammonia, methane, and hydrogen. For about a week Miller and Urey allowed electric charges to go through this mixture of gases. As the water vapor condensed, organic molecules—formed in the gaseous phase—began to accumulate.

Figure 6
Stanley L. Miller.

Thus Oparin's ideas on the origin of organic molecules, and their possible prebiotic syntheses, were successfully put to the test for the first time. In the years since the Miller-Urey experiment, many of the organic "raw materials" of contemporary cells have been synthesized in the laboratory. (Some examples are shown in table 1.) Hydrocarbons and fatty acids form from carbon monoxide (CO) and hydrogen as shown in reaction 1, for example. Small amounts of ribose and deoxyribose, the sugars found in nucleic acids, can be formed from formaldehyde and acetaldehyde. It is now possible to form nonenzymatically the amino acids (reactions 4 and 5) that make up the proteins in all living beings. As was brilliantly shown by Juan Oró, we can form adenine from hydrogen cyanide, and other simple "prebiotic" reactions lead to the synthesis of guanine, uracil, cytosine, and thymine (reactions 6, 7, and 8). Some of these molecules even condense together to form oligomers or polymers. Under the appropriate conditions, amino acids form oligopeptides and nucleotides form oligonucleotides. Finally (reactions 11 and 12), we can nonenzymatically form the phospholipid components found in all cell membranes. Experiments of this kind suggest that Oparin and Haldane were on the right track. The chemical syntheses of organic compounds on the prebiotic Earth probably yielded a mixture of many different kinds of molecules, including many of those found in contemporary cells.

Table 1
Major prebiotic reaction pathways.

1.	CO + H$_2$	FTT (Fischer-Tropsch type) catalysis	Hydrocarbons, fatty acids
2.	CH$_2$O	Base catalysis	Ribose
3.	CH$_3$CHO CH$_2$O	Base catalysis	Deoxyribose
4.	RCHO HCN NH$_3$	Strecker condensation	Amino acids, hydroxyacids
5.	As above + CH$_2$S	Strecker condensation	Cysteine, methionine
6.	HCN	Base catalysis	Adenine, guanine
7.	HC$_3$N + urea	Condensation	Uracil, cytosine
8.	As above + CH$_2$O	Hydrazine	Thymine
9.	Amino acids	Cyanamide	Oligopeptides
10.	Mononucleotides	Cyanamide	Oligonucleotides
11.	Isoprene	Ultraviolet, ionizing radiation	Polyisoprenoids
12.	Fatty acids, glycerols, phosphates, bases	Cyanamide	Neutral lipids, phospholipids

Proteins or DNA First?

The problem, of course, is the transition from the mixture of organic compounds we think was present on the primitive Earth to living and reproducing organisms. Two sets of properties distinguish living from nonliving systems: *autopoiesis* and *reproduction*. Autopoiesis, a term derived from Greek, means "self-making." As defined by the Chilean biologists Francisco Varela and Humberto Maturana, it refers to entities that are separated from their environment by an interface or membrane and that *metabolize* (i.e., chemically maintain and perpetuate their identity in fluctuating environments). Autopoiesis is considered a prerequisite for reproduction. All contemporary cells have their

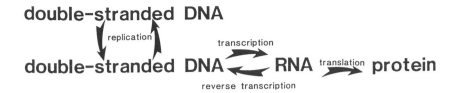

Figure 7
The central dogma of biology: DNA is transcribed to RNA, which is then translated into proteins. Reverse transcription allows RNA to be transcribed back into DNA.

genetic information stored in the form of double-stranded DNA. This DNA can replicate itself to produce more DNA; it can also give rise to RNA in a process known as *transcription*. Transcribed RNA contains information derived from the double-stranded DNA. In the process of translation, the information contained in transcribed RNA molecules is used on ribosomes in stringing together amino acids into proteins. Transcription and translation seem to be universal in all contemporary cells. This poses the problem of how such complicated systems arose from the disaggregate mixture of organic molecules on the primitive Earth.

Many people have compared the problem of the origin of life to the chicken-and-egg problem: Which came first, DNA or proteins? On my view neither DNA nor proteins came first. Along with many others, I am convinced that RNA came first. Like DNA, RNA is a molecule that can carry genetic information. Like proteinaceous enzymes, RNA can act as a catalyst. Unlike proteinaceous enzymes or DNA, RNA is autocatalytic. RNA's catalytic activity may be understood as a remnant of times in the prebiotic environment when these molecules were surrounded by lipid membranes.

If RNA came first, this question arises: Why was there a transition in living beings from information stored in RNA to information stored in the form of double-stranded DNA? Several scientists have come to the following conclusions:

• The backbone of single-stranded RNA is much less stable than the equivalent structure in double-stranded DNA. Enhanced stability gives DNA greater fidelity as an information store.

• RNA is much more susceptible to chemical transformations, such as the deamination of cytosine into uracil.

• RNA molecules absorb more ultraviolet radiation than double-stranded DNA, so the chances of the genetic information being damaged by the UV radiation on the primitive Earth were greater for RNA than for DNA.

• RNA polymerase, the enzyme that forms RNA from a DNA template, does not have a proofreading activity. Whereas DNA has many repair mechanisms, RNA has none.

To many, these facts suggest that the evolution of RNA to DNA took place very quickly. Of course these ideas will be developed further, just as the ideas of Herrera, Haldane, Oparin, and Urey have been refined and rewarded. But we can also be absolutely sure that some of these assertions will be abandoned for a more complete idea of the way life appeared on Earth.

Readings

Deamer, D. W., and G. R. Fleischaker. 1994. *Origins of Life: The Central Concepts.* Jones and Bartlett.

Oparin, A. I. 1953. *The Origin of Life*. Second edition. Dover.

Oró, J., S. L. Miller, and A. Lazcano. 1990. The origin and early evolution of life on Earth. *Annual Review of Earth and Planetary Sciences* 18: 317–356.

Urey, H. 1952. On the early chemical history of the Earth and the origin of life. *Proceedings of the National Academy of Sciences* 38: 351–363.

Varela, F., and H. Maturana. 1974. Autopoiesis: The organization of living systems, its characterization and a model. *BioSystems* 5: 187–196.

6 Clues to Life in the Archean Eon

Paul K. Strother and Elso S. Barghoorn

Most scientists accept the concept that life originated on Earth at the beginning of the Archean eon, at least 3.4 billion years ago. But upon what is this common belief based? In this chapter Paul Strother and Elso Barghoorn present evidence of the earliest life collected from widely separated ancient fossil-bearing localities. Chemical, contextual, and comparative scrutiny of these preserved microenvironments and of extant bacterial microcosms allow Strother and Barghoorn to help resolve the true nature of the oldest putative microfossils.

One could argue that the study of the origin of life as an event is more a philosophical than a scientific pursuit. The fact of the origin of life we can accept, but the direct documentation of an origin "event" in Earth history is probably not possible. We can, however, document a pattern of early evolution based on the recovery of an imperfectly preserved fossil record. The earliest fossil record is very different from the bones, shells, and coals of the traditional paleontological trade. Instead, we must search for either the direct organic remains of microscopic life, or, indirect evidence of the activity of entire communities of microorganisms, such as stromatolites. The collective record of this direct and indirect evidence for the earliest ecosystem indicates that microbial communities inhabited shallow marine waters beginning about 3.4 Gyr ago. Such a conclusion remains somewhat speculative and is colored by a complex set of arguments concerning the biogenicity of virtually every claim of a fossil find from the Archean Eon.

The purpose of this chapter is to now take a look at the Archean record, both pro and con. We will approach this review by first looking at two deposits of Early Proterozoic age (2.0 Ga)*: the Gunflint Chert

---

*Ga: billions of years ago.

and the Duck Creek Dolomite. Since the biogenicity of microfossils from these deposits is not disputed, they give us a basis by which to investigate older, and more problematic, Archean assemblages. The sources of information on this subject are widespread and include geologic formations on all the continents (figure 1).

Early Proterozoic Microfossils

The formation, aptly named the Gunflint Iron Formation, is found in northern Ontario and adjacent Minnesota and consists of a series of sediments including a thin but persistent unit of rocks called cherts or flints. Sparks from chert-struck steel ignited gunpowder in the flintlock rifles of yesteryear. These cherts contain large numbers of microorganisms still relatively unaltered in shape and preserved in life position. The Gunflint microbiota, as we call it, contains several different kinds of microbial communities preserved in both flat-laminated and domal stromatolitic cherts.

Most Precambrian rock samples are studied by making very thin slices, rendering them translucent to light, and observing them microscopically. Objects in these petrographic thin sections are indigenous to the rock sample itself, and microfossils observed this way were necessarily incorporated into the rock at the time of its formation. Contamination of more recent material may occur where there are either microcracks or spaces between individual mineral grains that make up the composite rock. But, in the Gunflint Chert samples, the sheer abundance of microfossils and their relation to the internal mineral fabric of the rock leave no doubt as to their indigenous origin.

A thin section of the Gunflint chert, observed under the microscope, contains a profusion of filaments (*Gunflintia*) and small ovoid bodies (*Huroniospora*) very similar in appearance to bacteria that you might typically find in pond scum (figure 2). The overall Gunflint assemblage consists of vast numbers of individual microorganisms that have been placed into perhaps two dozen different genera. The fossil microbes are morphologically complex and diverse; they include complicated, rosette-shaped (*Kakabekia*) and multi-layered budding forms of bacteria (*Galaxiopsis*) (figure 3). This morphological complexity and diversity leads to the universal acceptance of the Gunflint microfossils as bona fide examples of microbial life at 2 Ga.

The Gunflint microbiota is not an isolated quirk of evolution, similar assemblages of about the same age have been found the world over. One such example comes from the Duck Creek Dolomite of Western Australia where fossiliferous cherts are found mixed in with stromatolitic and oncolitic carbonates. The Duck Creek assemblage is especially important in our discussion of the biogenicity of more ancient (pre-2 Ga) deposits, because it contains a natural experiment in fossilization. Cherts from the Duck Creek Dolomite contain microfossils in different stages of degradation prior to complete fossilization. Consequently they reveal populations of carbonized spheres which are poorly preserved, yet clearly biogenetic, based on their association with well-preserved populations within the same sections (figure 4). These biogenic populations of poorly preserved spheres are indistinguishable from some much older samples. Also, and perhaps more importantly, individual, isolated filaments are found in the cherty matrix between organic-rich portions of the chert (figure 5). These filaments or sheaths probably represent the last colonizers of degraded cells prior to fossilization. They document a model system whereby isolated, individual sheaths can be preserved in cherts without being part of a much larger population of similar cells. As we shall see in the next section, this model can be used to lend support to the claims of biogenicity for some isolated "filaments" that represent the oldest fossils on Earth.

Biogenicity Criteria

Two types of rock sequences or terrains occur in the Archean Eon: high-grade metamorphic terrains and greenstone belts. The high-grade metamorphic terrains, by volume, are more significant; they consist mostly of granites and gneiss. These metamorphic rocks represent the primordial fractionation of the upper mantle into crust added to partially melted surface materials. The greenstone belts are volcaniclastic and sedimentary sequences which have been subsequently metamorphosed to produce a suite of minerals, including chlorite, that are characteristically green. These rocks are metasediments—sedimentary rocks that are slightly metamorphosed but retain some of the properties, such as texture, that were present when they were unconsolidated sediments. Metamorphism here refers to the processes of increasing

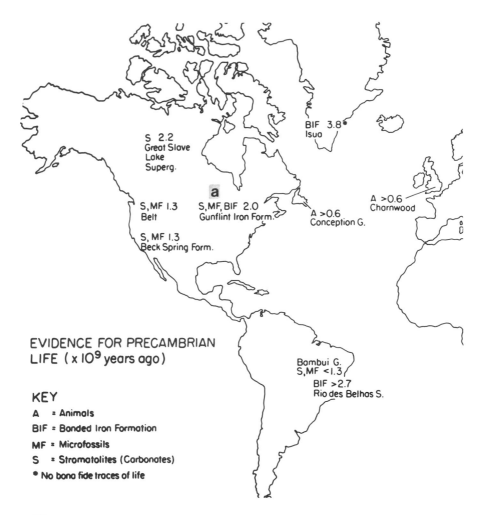

EVIDENCE FOR PRECAMBRIAN
LIFE (x 10⁹ years ago)

KEY

A   = Animals

BIF = Banded Iron Formation

MF = Microfossils

S   = Stromatolites (Carbonates)

* No bona fide traces of life

Figure 1
World map showing evidence of Precambrian life: (a) Gunflint Iron Formation in Ontario
and Minnesota, (b) Tree Formation in Swaziland, (c) Bitter Springs Formation in central
Australia. *(drawing by J. S. Alexander)*

A >0.6
Valdai Se.

Sulfur Springs
S, MF 1.1

Krivoy Rog
BIF 2.1

Jixian
S, MF 1.9
S, BIF >1.7
Changzhougou
Form.

S, BIF 2.6
Dharwar Aravalli G.

Swaziland Superg.
MF,S b 3.4
A >0.6 S, MF, BIF 2.3
Nama Sy. Transvaal Superg.

Warrawoona G.
S, MF 3.5 C
S, MF 2.8 S, MF 0.9
Fortescue G. Bitter Springs Form.
A >0.6/
Pound Quartzite

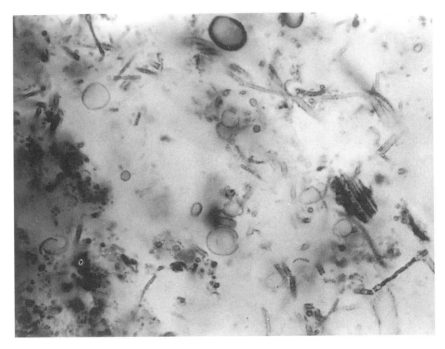

Figure 2
A typical thin section of microfossils from the 2-Ga Gunflint Chert reveals a mixture of
spherical cells (*Huroniospora*) and thin bacterial filaments (*Gunflintia*). The miscellaneous
orientation of these cell populations mimics that seen in modern pond scum where
heterotrophic bacteria dominate the degradation of organic matter. Magnification: approximately 500×.

heat and temperature that occur with burial during formation of sedimentary sequences over time, rather than the high-grade metamorphic
changes that occur when rocks reach a molten or partially melted state.
Because these two general rock types have been metamorphosed,
however, the veracity of almost all fossil occurrences from this time
period is under question to some extent.

Direct remains of life in Archean rocks consist of degraded remains
of individual cells with simple morphologies. The likelihood that a
sample of such simple structures is of biological origin must be evaluated using all the available evidence associated with that particular
sample. If the fossil consists of a group of cells rather than isolated
single cells, the likelihood that the sample has a biological origin
increases. Cells found grouped in clusters or in populations that can be
described statistically are more likely to be of biological origin.

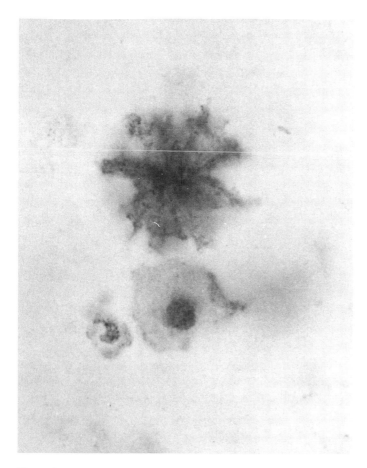

Figure 3
A thin section of two of the typical microfossils found in the "Frustration Bay" Facies
of the Gunflint Chert. The upper form (*Kakabekia*) consists of thin organic strands that
radiate out from a central point and are covered by a thin membrane. It has been
compared to the modern Mn-oxidizing bacterium, *Metallogenium*. *Galaxiopsis*, a microfossil with no known modern counterpart, has a dark central body surrounded by one
or more globular layers. It is similar in shape to the living, thermophilic, sulfur-oxidizing
bacterium, *Sulfolobus*, but the fossil is many times larger in size. The *Galaxiopsis* and
Kakabekia morphotypes are sometimes found connected to each other. Magnification:
approximately 1500×.

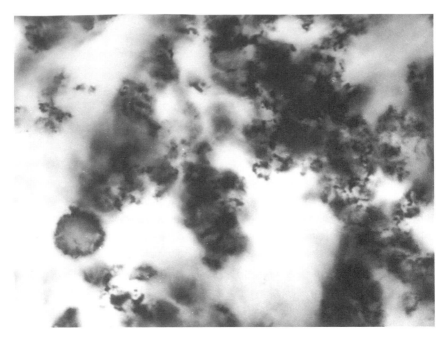

Figure 4
A single well-preserved spherical microfossil from the Duck Creek Dolomite, Western
Australia, found in close association with severely degraded forms, seen here as dis-
persed granular carbon. Magnification: approximately 1000×.

Complex, biological morphology increases the likelihood that a microfossil is real. Putative fossils composed of simple spheres are very
difficult to identify. Many different microorganisms appear spherical.
However, if the morphology of the microfossil in question is filamentous, colonial, or even more complex, the possibility that the sample is
indeed biological is greater.

Morphological features can be used to indicate contamination by
more recent forms. The surface texture of individual microfossils tends
to be highly granular. Smooth and intact spheres in a rock sample are
often suspect. Comparison with degraded organic matter found within
a particular deposit may be of value in assessing biogenicity. Color is
also an important criterion for assessing contamination. Organic materials 3 billion years old are usually entirely carbonized and appear
black in thin section when viewed through a transmitted-light microscope. Brown or yellow indicates organic materials that have not been

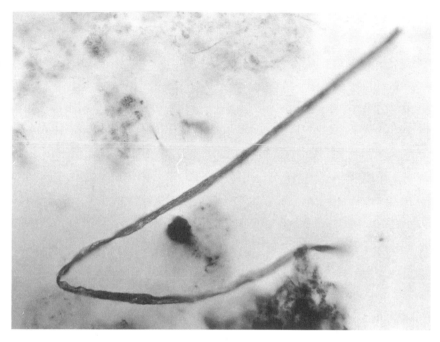

Figure 5
A single well-preserved filamentous microfossil from the Duck Creek Dolomite, found in close association with dispersed granular carbon which does not contain apparent microfossils. Well-preserved individuals within a cherty matrix dominated by severely degraded organic matter at 2 Ga present an important test case for the claims of biogenicity made for most microfossils greater than 3 billion years in age. Magnification: approximately 1000×.

fully degraded, implying contamination of some kind. This tendency of organic matter to change color over time is called organic maturation. Color changes from yellow to brown to black are accompanied by an increasing relative percentage of carbon. Over time, complex organic matter releases volatile components containing hydrogen, nitrogen, and oxygen from the original organic constituents. The carbon remains, increasing the relative percentage of carbon and the blackness over time as the organic material ages. From the color, we can roughly assess the extent to which the organic matter has been metamorphosed.

Another criterion of biogenicity is the sedimentary context of the microfossils. The sedimentary context of a particular assemblage of direct remains is inferred from the mineralogy of the sample; certain

Figure 6
Pillow lavas in the Warrawoona Volcanics. These lavas were extruded under water.
Their characteristic dome shape has been weathered smooth, leaving ovoid surfaces.

types of minerals are characteristic of particular depositional environments. The Swaziland Supergroup, for example, contains sediments
laid down under water, with ripple marks and other evidence of water
flow embedded within sandstones. Other examples of sedimentary
environments in Archean rock sequences include volcaniclastic deposition, ash flows, ash falls, and subaqueous submarine volcanic deposition as lava flows or pillow lavas (figure 6).

Analysis of Putative Microfossils

The Barberton Mountain Land in the eastern Transvaal of the Republic
of South Africa and Swaziland contains a series of rare sedimentary
rocks, volcanic rocks, and fluvial deposits thousands of meters thick.
The middle section of this series, the Fig Tree Formation or Fig Tree
Series, contains massive units of dense black cherts containing a relatively high concentration of organic matter. Closer examination of
these rocks in thin section reveals the presence of spherical microstructures similar to the degraded biogenic populations from the Duck

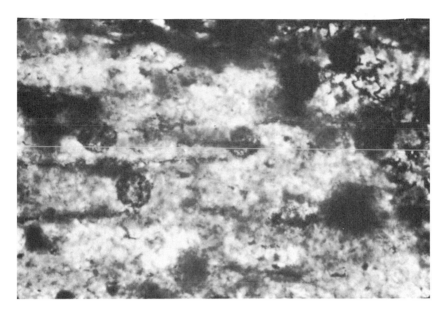

Figure 7
Archaeosphaeroides from the Tree formation.

Creek cherts (figure 4). These microstructures are clearly not mineral inclusions, and their organic composition can be readily demonstrated by examination or by simple combustion. The fossil cells are contained within the matrix of the rock and were present before the rock was silicified. What makes the microfossils of the Fig Tree rocks interesting to geologists is their great antiquity. The rocks of the Fig Tree system have repeatedly been dated to 3.4 Ga. Hence, the Fig Tree microbiotabiota is considerably older than the assemblage from the Gunflint Formation.

A thin section of a sample from the Barberton Mountainland contains examples of the Fig Tree (now called Swartkoppie) microspheres (figure 7). These were first described by E. S. Barghoorn and J. W. Schopf as *Archaeospheroides barbertonensis*. The microfossils are defined by their roughly spherical shape, carbon composition, and a coarse granular texture outlining their spherical shape. Populations of these microspheres occur in Archean cherts with organic laminae. Volcaniclastic textures are where black organic matter is present. The original organic material has been subsequently enclosed by microcrystalline silica, or chert. *A. barbertonensis*-type microspheres are black in a high-power light micrograph of the layer containing them (figure 7). The original sedimentary context is preserved as slightly undulating laminations. The isolated, perfectly spherical small entities have an average diameter of 18 microns and a surface texture that is granular and coarse.

The Barberton spheroids may occur either as isolated spheres or associated with one or more other cells. They are not particularly well preserved, which is expected of material of this antiquity. The spheres have the same morphology and surface texture whether solitary or clustered. Also, these spheres compare favorably with the degraded cells from the much younger Duck Creek Dolomite mentioned earlier.

In addition to the populations of *A. barbertonensis*, other populations of cell-like microspheres exist in cherts from the Barberton area. Size-frequency diagrams can be used to compare these "ghost" populations with the *A. barbertonensis*-type spheres (figure 8). These graphs can also be compared with similar graphs derived from living populations of spherical cells or from populations from abiogenically produced proteinoid microspheres created in the laboratory. The mean diameter of the microspheres can be compared, as well as the flatness or the shapes of the histograms. In the upper graph of figure 8 the mean diameter is skewed slightly to the left. In the lower graph the mean diameter is skewed little, if at all, and the histogram is fairly uniform. This suggests that, even within the Barberton material, multiple populations of differing origin are present. In this case, the "ghost" population could well be organic coatings of sedimentary grains rather that the remains of intact cells

Sidney Fox and his colleagues at the University of Miami have set up an experimental system in which they create microspheres composed of protein-like substances thought to be readily available on the pre-biotic Earth. Their proteinoid microspheres (coacervates—see chapter 5) represent a possible type of protocell. But with such simple spheroidal morphology, it is difficult to ascertain whether the microspheres we see in the Swaziland and Warrawoona cherts were derived from once-living cells, protobionts, or proteinoid microspheres.

The Warrawoona Volcanics

Another microfossil-containing environment is represented in rocks about 3.4 billion or 3.5 billion years old from the Pilbara Block in

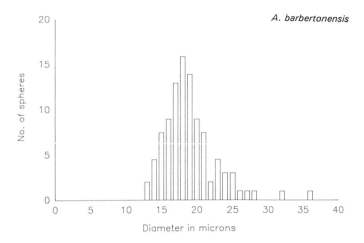

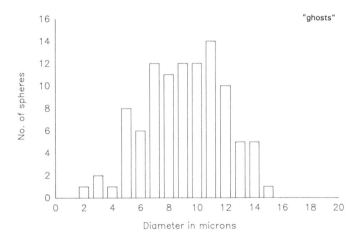

Figure 8
Size-frequency histograms of microsphere populations. The upper graph shows the distribution of diameters of *Archaeospheroides barbertonensis* microspheres. The lower graph plots the frequency of sizes of spheroidal "ghost" microfossils.

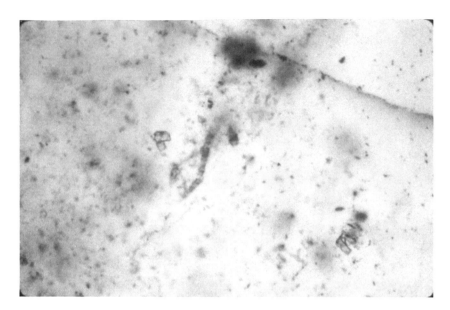

Figure 9
Photomicrograph of thin section from Warrawoona Volcanics, showing septate filaments.

the Warrawoona Volcanics of Western Australia. Figure 9 shows a
filament from the banded Warrawoona chert. The presence of
filaments rather than solitary cells indicates a higher likelihood of
biogenicity. An even more complicated fossil from another Warrawoona sample is shown in figure 10. Thin filaments radiate from the
dark clustered area near the center. This rosette is so complex that it is
difficult to attribute it to an abiogenic source. Such rosette morphology
can be seen in modern bacteria. Other rocks from the Warrawoona
area contain evidence of shallow marine and possibly evaporitic
conditions. This gives a reasonable setting to the Warrawoona microfossils.

It is now generally accepted that filaments (sheaths) similar to those
in figures 9 and 10, from the Warrawoona Volcanics represent the
oldest fossils on Earth. These fossils are found as isolated occurrences,
not as large populations, so the model represented by the isolated
filaments from the younger Duck Creek Dolomite is an important one
for the interpretation of these most ancient structures.

Different Kinds of Archean Fossils

Evidence for Archean life ranges from the direct remains of cells and their degraded carbon-containing molecular constituents to microbial laminates and stromatolites, which only indicate the activity of former organisms. Some individual kinds of mineral grains such as apatite (a phosphate-containing mineral) or magnetite (an iron oxide) may be uniquely associated with organisms. The recovery of such mineral grains has been used to promote the idea that living systems existed as far back as 3.85 Ga.

Both microbial laminates and stromatolites are sedimentary rocks in which the fabric and geometric relationships of the layers, as well as the mineral composition, are mediated by the activity of microorganisms (usually, cyanobacteria). Communities of microorganisms trap and bind sediment to form distinctively laminated rocks. Stromatolitic laminations are usually convex upward domes, whereas the laminations of algal laminates are nearly flat or wavy-bedded. The stratigraphic record of stromatolites begins around 3.0 to 2.7 Ga, although some rather doubtful forms have been described from older sediments.

Since living communities of bacteria and cyanobacteria form similar structures today, we suspect that Archean structures had the same origin. However, microlaminated structures can occur as part of sedimentary rocks in several ways that are not necessarily related to the trapping and binding of sediment by microbial ecosystems. Furthermore, stromatolites themselves only rarely contain fossils, since the organisms that construct them tend to degrade, leaving behind only layered rock devoid of fossils. Recent research by John Grotzinger and his colleagues at MIT has shown that some ancient "stromatolites" probably formed abiologically, in the absence of microbes as trapping and binding agents. Geologists will now have to examine more closely the stratigraphic record of stromatolites to assure that Archean stromatolites were biogenic.

Another source of information about Archean life is organic chemical fossils. Such organic residues retain no morphological structure of the original cell. Carbon-containing compounds in rocks, comprising complex organic molecules generated only by living systems, have been isolated. For example, porphyrin derivatives found in Archean rocks are evidence for the existence of chlorophyll, or at least of living systems, at the time. The clearest evidence of ancient porphyrins is from rocks about a billion years old. Gas chromatography of the extracts from the Gunflint Formation shows the presence of two probable degradation products of chlorophyll: the hydrocarbons phytane and pristane. Both are side chains on the central tetrapyrrole ring of chlorophyll molecules, whose release during degradation has been demonstrated in the laboratory. The presence of such compounds merely implies the existence of photosynthesis at this time, because, porphyrins can come from heme groups, and phytane and pristane can be lipid components of heterotrophic bacteria. However, no one doubts the biogenic nature of these complex biomolecule derivitives.

The origin of photosynthesis is a milestone in evolution whereby a continually available source of energy, in the form of sunlight, could now be converted to chemical energy and used in a myriad of biochemical and biosynthetic pathways. Hence, the antiquity of photosynthesis is fundamental not only for the establishment of primordial ecosystems but also for basic bioenergetics. Can we determine when photosynthesis began on Earth? It so happens there is a chemical mechanism whereby some putative evidence can be extracted.

Carbon dioxide, one of the two fundamental raw materials (along with water) in the photosynthetic equation, exists in the atmosphere in the form of two stable isotopes: $^{12}CO_2$ and $^{13}CO_2$. (^{14}C, a radiogenic isotope, is an extremely minute constituent of atmospheric carbon dioxide.) The ratio of the two isotopes in the atmosphere is very disproportionate: 99 percent is in the form of the lighter form, $^{12}CO_2$, and about 1 percent in the form of $^{13}CO_2$. Both isotopes are drawn into the photosynthetic organism during photosynthesis. Because of the kinetics of the molecules during biochemical reactions, the lighter isotope, being of higher velocity, is selectively concentrated by organisms in the process of photosynthesis. This leads to a fractionation of the two isotopes, so the ratio of carbon isotopes incorporated into the organic material of the body is not in equilibrium with the isotopes of carbon existing in the atmosphere. In other words, the ratio of ^{13}C to ^{12}C in the organism is lower than the ratio of ^{13}C to ^{12}C in the atmosphere. By comparing the ratio of the isotopes against the standard ratio in the atmosphere, one can discriminate between inorganic carbon and carbon formed by photo- and lithotrophic carbon fixation. Analysis of organic carbon from shales suggests the presence of this isotopically light, photosynthetic carbon as far back as 3.4 Ga.

A problem with chemical fossils is that all rocks are permeable to some extent; over long periods of time, chemical contamination is likely. Indeed, claims as to the indigenous nature of organic matter from the 3.8 Gyr old Isua Suprecrustal Complex have been shown to be contamination from modern lichens.

Precambrian Micropaleontology and the Origins of Life

What does Precambrian micropaleontology have to tell us about the framework of research on the origin of life? The first organisms arose, we can assume, from inorganic sources, but in order to produce life there must previously have been some aggregation of simple materials into far more complex organic compounds. Hence, we can postulate major sequences of events between the origin of the Earth and the origin of life as a system of syntheses in which prebiotic or chemical "evolution" was taking place in organism-free molecular systems. Such postulates are the legacy of Oparin, who brought a biochemical view of abiogenesis into the realm of scientific study.

The fossil record has no direct bearing on the study of the processes that led to the origins of the earliest terrestrial ecosystem. But the subsequent effects of life activities and metabolic production are an essential part of the cycling and distribution of elements on the Earth's active surface. The fossil record of early life and its associated chemical and trace fossils are extremely important in determining the timing and sequence (i.e., the pattern) of evolutionary events that can only be speculated upon by biologists today. Evolution does not always appear to fit a simple progressive or cumulative sequence—we can expect to find many glitches and surprises about early evolutionary patterns as more researchers pursue the study of Archean life.

In conclusion, working with Archean materials is still very problematic. Material is rare, only a few unmetamorphosed sedimentary rock types are contained within the greenstone belts. The presence of unmetamorphosed or only mildly metamorphic sediments is an absolute prerequisite for finding legitimate organic fossils. Furthermore, not all laminated rocks are stromatolitic, even though they have an undulating texture. The contortions and microlaminations of sinter and siliceous rocks that form by precipitation from high temperatures can resemble an algal laminate or a stromatolite. Sinter is composed of opal (silica) from hot spring water with high mineral concentrations. The genesis of sinter is entirely abiological; it is not related to stromatolites and algal laminates, which form, by definition, by a combination of biological and sedimentary processes.

Certain criteria can be used to establish the biogenicity of a particular laminate sample. In the sedimentary context, one must follow the laminae laterally and vertically to determine the strata to the sides of and immediately above and below a presumed stromatolitic horizon. The fabric of the laminate itself must be carefully analyzed; the type, texture, and orientation of the mineral grains making up the laminae must be considered. Extant microbial mats and stromatolites display only certain fabrics. Careful comparison should be made between the microlaminated rocks in question and materials known to be biogenically derived.

The most reliable Archean fossils are direct organic remains which come from measured stratigraphic sequences within a known sedimentological context. The best evidence of biogenic remains comes from samples containing significant populations of unmistakable fossils. Schopf's discovery of Western Australian microfossils from the "Apex Basalt" chert is a case in point. The filaments are probably remains of cyanobacteria. When the ancient fossils are abundant and recognizable, comparisons can be made between Archean and other populations, living and fossil, to help determine with confidence the nature of the clues to life in the Archean Eon.

Readings

Barghoorn, E. S. 1971. The oldest fossils. *Scientific American* 244, no. 5: 30–41.

Barghoorn, E. S., and S. A. Tyler. 1965. Microorganisms from the Gunflint chert. *Science* 147: 563–577.

Buick, R. 1990. Microfossil recognition in Archean rocks: An appraisal of spheroids and filaments from a 3,500 m. y. old chert-barite unit at North Pole, Western Australia. *Palaios* 5: 441–459.

Engel, M. H., S. A. Macko, and J. A. Silfer. 1990. Carbon isotope composition of individual amino acids in the Murchison meteorite. *Nature* 348: 47–49.

Knoll, A. H., and E. S. Barghoorn. 1977. Archean microfossils showing cell division from the Swaziland System of South Africa. *Science* 198: 396–398.

Schopf, J. W., ed. 1983. *Earth's Earliest Biosphere: Its Origin and Evolution.* Princeton University Press.

Schopf, J. W. 1993. Microfossils of the Early Archean Apex chert: New evidence for the antiquity of life. *Science* 260: 640–646.

Tyler, S. A., and E. S. Barghoorn. 1954. Occurrence of structurally preserved plants in pre-Cambrian Rocks of the Canadian Shield. *Science* 119: 606–608.

7 Microbial Landscapes: Abu Dhabi and Shark Bay

Stjepko Golubic

Most of the evidence of life during the Archean and Proterozoic eons is in the form of stromatolites, which are fossilized microbial mats. In this chapter, Stjepko Golubic introduces these microbial landscapes. What is the role of live bacteria in the making of carbonate rocks? Golubic describes the formation of microbial mats and stromatolites at two widely separated locations: Abu Dhabi, on the Arabian/Persian Gulf coast, and Shark Bay, in Western Australia. After reviewing the main features of the mat communities and their constituent micro-organisms, he examines their activities in the formation of live mats and salt flats and their fossil remains. His discussion addresses ranges of dimension from the microscopic to entire landscapes.

Ancient Microbial Landscapes: Clues to Planetary History

Most of our knowledge of the Earth is limited to the last fifth of its history; we know very little about the first four-fifths, the Precambrian era. The dominant forms of life during these first four-fifths of our planet's history were microbes with prokaryotic cell organization. Prokaryotes, which include all bacteria, are membrane-bounded cells that lack membrane-bounded nuclei; eukaryotes are composed of cells that contain nuclei. Most eukaryotes also contain other membrane-bounded organelles, such as mitochondria. In examining the ancient geological record, one frequently encounters peculiar laminated rocks called *stromatolites*. These were formed by communities of live organisms, especially those dominated by cyanobacteria (photosynthetic oxygen-producing prokaryotes). In the Archean eon, Cyanobacteria were the dominant primary producers of organic matter on the planet, able to use solar energy and mineral (inorganic) matter as nutrients.

Organisms that turn inorganic nutrients into organic matter are called *autotrophs* (self-feeders). Plants and algae are the major primary producers today. Cyanobacteria constitute a minor component of modern ecosystems, although these resilient microorganisms often prevail when conditions become extreme. Without autotrophs as the primary producers, life would not be possible. Heterotrophs (organisms that feed on organic matter) depend utterly on autotrophs for food and energy.

Cyanobacteria and other microorganisms associated with them form tightly interwoven microbial mats. The term "microbial mat" refers to entire ecosystems of microorganisms. These structures used to be called "algal mats," but ultrastructural studies show that they are composed primarily of bacteria. The term "alga" (literally "water plant") is today restricted to eukaryotic oxygenic photosynthesizers belonging to the kingdom Protoctista. As in other ecosystems, the energy and nutrients are transferred from primary producers to consumers, including decomposers. While nutrients can be recycled, the energy passes through the system. Because energy transfer in all ecosystems follows the laws of thermodynamics, only a fraction of the energy captured from the sun is transferred to the heterotrophs.

Fossil stromatolites—solid, layered rocks composed of alternating organic-rich and mineral-rich laminae—provide a cumulative record of mat activities. During the vast eons of the Archean and the Proterozoic, living stromatolitic structures of varying sizes and shapes inhabited the seas and lakes, formed reefs, coated mud flats, and encrusted land.

Besides the study of stromatolites, another approach to Archean and Proterozoic microbial ecosystems exists: the study of modern environments as models of Precambrian conditions. In the few such environments we know of today we study the microbial and biogeochemical interactions that occur at the sediment-water interface. By comparing fossils and living organisms we learn that ancient stromatolites were built by microbes through the trapping and binding of sediment particles, often followed by mineral precipitation which solidified these structures. We know that the bacteria must have been able to stabilize sediment surfaces by the formation of mats. They recolonize the surface by gliding movement and growth, escaping total burial by sediment. The laminated structures we find today originated in the alternation of mat burial by sediment with periods of reestablishment of microbial mats on the surface of the deposited sediment. Our first example that offers an opportunity to study living stromatolites is in Abu Dhabi on the Arabian/Persian Gulf coast, where the activities of microbial mats have led to the formation of salt flats.

Microbial Mats of Abu Dhabi

A coastal strip in the southernmost part of the Arabian/Persian Gulf, near Abu Dhabi (figure 1), reveals a lagoon environment separated from the land by a wide salt flat (called a *sabkha* in Arabic). Between the lagoon and the sabkha is the darker intertidal zone of microbial mats. This zone acts as a factory that, over time, creates the large flat landscape of the sabkha.

The intertidal zone, the area between the water and the land, is a very harsh environment for life. During high tide the entire zone is flooded; during low tide the seawater retreats as the area is exposed to air and brutal dryness. Along the coast the high air humidity causes dew, yet only a few centimeters of rain falls every 3–4 years. Evaporation at low tide concentrates the dissolved salts, which pass through saturation stages and then precipitate. This process, a regular sequence of mineral formation, starts with deposition of calcium carbonate ($CaCO_3$), followed by gypsum ($CaSO_4 \cdot 2H_2O$) and halite ($NaCl$). The salinity of the lagoons and tidal waters is always above that of the normal seawater. Tidal fluctuations expose the organisms to the severe shock both of the presence of the water and its high ionic strength. Some microbes retain water in their cells and tissues so that their physiological activity continues during low tide whereas others become dormant. This coast is also exposed to high-intensity sunlight, which affects the mats differently during high and low tide. The distinctive zonal distribution of mat types results from variation in air exposure, desiccation, solar irradiation, and the effects of water.

Desiccation and solar irradiation are among the most important factors exerting selective pressures on the organisms. The upper intertidal zone is drier than the lower. Organisms less sensitive to desiccation live higher up in the intertidal zone, where they escape

Figure 1
Aerial view of sabkha near Abu Dhabi, United Arab Emirates. This lagoonal environment is really a vast microbial landscape. The waters of the Arabian/Persian Gulf are visible at the top of the photograph, and the land at the bottom. The darker areas in the middle constitute the intertidal zone, dominated by the microbial mats that are actually creating the sabkha landscape.

competition with those bound to the subtidal and lower intertidal regions. Greater microbial diversity and larger populations are seen in the lower zones.

Very few halophytes, salt-tolerant vascular plants, or other eukaryotes can even venture close to this harsh environment. The major primary producers are photosynthetic prokaryotes, for the most part cyanobacteria. Moving from the lagoon across the microbial belt of the intertidal zone toward land, one crosses a sequence of zones covered by different types of cyanobacterial mats, designated here (figure 2) as types a–f. These replace each other over a distance of 2–3 kilometers.

a. Gelatinous laminated biscuits are finely laminated structures formed by the cyanobacterium *Phormidium hendersonii* (figure 3). A small filamentous cyanobacterium, slightly more than a micron wide, is surrounded by a thin tubular polysaccharide sheath within which it

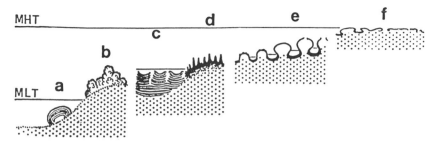

Figure 2
Microbial mat types found in the sabkha. A transition from subtidal (a) to supratidal (f)
mat types occurs over a distance of 2 or 3 kilometers. The various mat types are described
in the text. (a) Laminated biscuits. (b) Mamillate mat. (c) Low flat mat (*Microcoleus* mat).
(d) High pinnacle mat. (e) Convoluted mat. (f) Folded mat.

glides. Type a forms soft, gelatinous colonies scattered in the subtidal
portion of the lagoon. There are no coherent mats in that zone.

b. Mamillate mat is built by *Entophysalis major*, a coccoid cyanobacterium that occupies the lowest intertidal ranges. It is the first colonizer
of "mega-ripples" in the lower intertidal zone. These large sand rip-
ples, up to a meter high and several meters long, are formed in lagoons
and channels by currents that frequently change direction. The mamillate mat on these mega-ripples is a translucent, brownish, gelatinous
mass. The word *mamillate* is from the Latin *mamilla*, meaning "small
wart," and refers to the surface texture of the *Entophysalis* mat, which
is characterized by millimeter-size warts.

As we move further landward across the intertidal zone, the somewhat wavy landscape is completely covered by mats. This landscape
is differentiated into depressions which retain pools of stagnating
water during low tide and are covered by the low flat mat, and slightly
elevated, well-drained heights between them that are covered by the
high pinnacle mat. These mat types are formed not by populations of
individual species but by complex communities.

c. Low flat mat lines the bottom of pools and tidal channels in the
mid-intertidal zone. Low flat mat is a coherent, smooth, leathery microbial mat that in cross-section appears layered. This layering reveals

Figure 3
Laminated biscuit formed by the cyanobacterium *Phormidium hendersonii*. The small, soft, biscuit-shaped colonies, approximately 2–5 cm across, dot the sandy bottom of the near subtidal zone.

a vertically differentiated microbial community in which each layer constitutes a separate microenvironment that is dominated by the organism best suited to live under the prevailing conditions. The surface layer of the low flat mat is dominated by the filamentous cyanobacterium *Lyngbya aestuarii* (figure 4). Tightly packed discoid cells are surrounded by a firm, brown, gelatinous sheath. The extracellular gel of these sheaths contains the pigment scytonemine. Underlying *Lyngbya* is the *Microcoleus chthonoplastes* layer. This filamentous cyanobacterium forms bundles. Although *Microcoleus* produces sheaths, it does not produce extracellular pigments within them; it solves the problem of solar radiation by living beneath a protective layer of *Lyngbya aestuarii*. *Microcoleus* tolerates low oxygen tension, and thus colonizes stagnant pools, characteristically spreading horizontally along their floors. Below *Microcoleus* is a layer devoid of oxygen. Anoxygenic photosynthetic bacteria thrive here. Some are filamentous, similar to *Chloroflexus*, known in hot springs. Others are coccoid, motile and nonmotile purple sulfur bacteria. Below the purple zone is a black

Figure 4
Photomicrograph of *Lyngbya aestuarii*. This filamentous cyanobacterium is surrounded by a thick, gelatinous sheath and is found with *Microcoleus*, another cyanobacterium, in the low flat mat.

layer stained by ferrous sulfide (FeS). Here dwell sulfate-reducing bacteria. At this level and below grow heterotrophic anaerobic bacteria, including fermenters and methanogens. They enhance decomposition of organic matter produced by the mat photosynthesizers.

d. High pinnacle mat covers the elevated hills between pools, channel levees, and their well-drained slopes. The name comes from the growth of numerous upright pointed cones, or pinnacles, 1–2 cm high. As in the low flat mat, the surface layer of this mat is composed primarily of *Lyngbya aestuarii*. However, the principal organism in the main underlying layer is *Schizothrix splendida* (figure 5) rather than *Microcoleus chthonoplastes*. On well-drained sites *Schizothrix splendida* receives the ample water at high tide and the regular oxygen supply at low tide it requires. Below the *Schizothrix* resides a thin, often incompletely developed layer of purple photosynthetic bacteria, and an equally faint black layer, indicating a transient anaerobic zone. Beneath the high pinnacle mat are light-colored, oxidized sediments

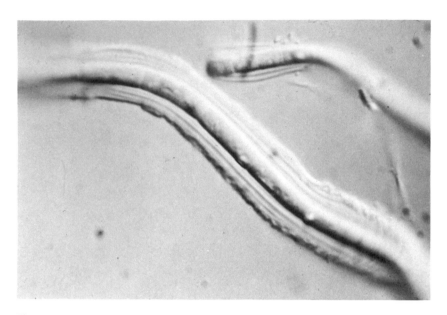

Figure 5
Photomicrograph of *Schizothrix splendida*. This sheathed cyanobacterium is a major component of pinnacled mat.

in contrast to the extensive purple layer (up to a centimeter thick) of the low flat mat, usually with massive, black, organic-rich anaerobic sediments.

e. Convoluted mat is a contiguous leathery mat that is characterized by a folded surface which forms over gas bubbles trapped below the mat. The dominant mat form in the upper intertidal zone, it combines elements of the low flat mat and pinnacle mat, but with different spatial distribution of the microorganisms. Lower portions contain primarily *Microcoleus chthonoplastes*; *Schizothrix splendida* are in the upper parts. Continuing growth increases the mat surface laterally. Convolutions that become more and more elaborate arise, increasing to about 10 centimeters in diameter. The lower, upward-concave folds hold water; the convex domes are hollow. The convolutions, thick in the lower portions and thin on top, often become perforated. This mat develops at a level where water trapped between tides is long exposed to evaporation. Gypsum precipitates from this highly concentrated brine, both in the mat's surface depressions and within the enclosed bubbles.

f. Folded mat is the final stage in the development of the intertidal microbial mats. With prolonged exposure in the uppermost ranges of the tidal flux, the entire mat periodically dries out. *Microcoleus* is replaced by *Schizothrix* throughout the mat. Dry conditions slow growth at the mat surface. Underneath the surface is an environment with enough moisture for degradation to continue. The balance between growth and degradation keeps the mat thin. As the mat gets thinner, the convolutions collapse and flatten, and the mat shrivels and becomes patchily disrupted. During low tide the mat dries out completely and becomes mummified. Ultimately, desiccation stops both constructive and destructive processes. The shriveled mat residue is blown away by wind.

Many unmentioned microorganisms are present in each mat type. Whereas mat types a and b are largely products of single types of cyanobacteria, interacting kinds of cyanobacteria form the complex microbial mat communities in c–f.

How do the main participants in the Abu Dhabi intertidal landscapes "drama" interact? After we consider dynamic processes that occur across the tidal flat, we can assess the changes in the landscape.

There are several rules: (1) Each mat organism has a set of specific requirements and responses. (2) Each naturally grows to occupy areas where its conditions are met. The cyanobacterium may (3) enhance its own position by optimizing the conditions or (4) may change the conditions in a direction that inhibits its own functions (while it creates opportunities for other mat inhabitants). Relationships between component mat organisms may be complementary, antagonistic, or neutral.

Building on Sand

The subtidal lagoon environments are dominated by physical forces that are the major shapers of sediment. Microbial growth is kept in check by animal bioturbation, especially by cerithid snails and worms that graze and churn through the sandy sediments. Solitary gelatinous biscuits of *Phormidium hendersonii*—perfect models of stromatolitic architecture—persist for periods of 30–60 days before they fall victim to the rapid turnover of organic matter. By day the gliding movement of *Phormidium* is positively phototactic, the filaments orient upward; by night it changes direction into a horizontal "resting" position. As it glides, *Phormidium* abandons its sheath and weaves a structure that each day is marked by a pair of laminae formed by the horizontally and vertically oriented sheaths. *Phormidium hendersonii* lives by a circadian rhythm and follows a solar clock. Evidence of its activity is ultimately obliterated; even a few months later nothing remains. The preservation potential is so low we cannot recognize ancient stromatolitic records of *Phormidium*. Yet from *Phormidium* growth patterns we learn principles that probably operated in the construction of countless ancient stromatolites.

The entire mid-tidal landscape is coated by coherent carpet-like mats. The first mat that colonizes loose sediment in the lower intertidal zone is the mamillate mat of *Entophysalis major*. This coccoid cyanobacterium excretes polysaccharide envelopes at the time of cell division. As a consequence, each cell is embedded in a cushion of gelatinous matter. Within these envelopes the microorganism deposits the extracellular yellow-brown pigment scytonemine. This light-induced pigment stains the surfaces of *Entophysalis* colonies, protecting them from excessive light. Just as ultraviolet radiation is detrimental to life, visible light, if too strong, can be detrimental. The sheath pigment scytonemine absorbs in both the ultraviolet and the visible range of the spectrum, although some ultraviolet radiation is also absorbed in the nonpigmented portion of the sheaths. Scytonemine pigment, resistant to degradation, preserves well. Similar pigments played a protective role in ancient mats and stromatolites.

When cores were taken from the oldest portions of the Abu Dhabi sabkha at a depth of 60–80 centimeters below the present sabkha surface, they showed distinct lamination. Microbial mats accrete at a rate of several millimeters per year. These 8000-year-old laminae contained cyanobacterial envelopes and sheaths with their brown pigment still preserved. Since scytonemine pigment develops only under the influence of strong direct sunlight, we can conclude that this portion of sabkha, now well below the surface, was once a surface mat illuminated by the sun. Pigmented sheaths remain preserved in fossils 1–2 billion years old, revealing the orientation of the mats with respect to the sun at the time they grew. In this way we use clues from modern environments to reconstruct ancient ones.

Trapping, Binding, and Polygonal Cracking

Each layer of a fossilized stromatolite represents an ancient mat, once active at the sediment-water interface, that was involved in trapping and binding sediment particles and/or promoting mineral precipitation. Sediment transport and deposition are essential components of the intertidal environment. Sediments, brought in by tidal currents, are dumped on or trapped in microbial mats. With every change of tide the bacteria must cope with rapid burial by sediments, and they have found two solutions to this problem: growing and moving. Some grow through the sediment fast enough to form a new layer on top of it by adding more cells to extend their filaments while one end of the filament remains anchored below. Many filamentous cyanobacteria employ a gliding movement to crawl out of their sheaths, which are left buried in the sediment while a new mat is established on the surface of the sediment. As sediment is deposited and the mat grows, the entire landscape becomes slightly higher with each tidal cycle.

Microorganisms trap sediments in several ways. Smaller sediment particles adhere to the surfaces of filaments and coccoids, which are very sticky because of the highly hydrated polysaccharide sheaths that most cyanobacteria produce. Mats also act as a filter mesh to trap larger sediment particles, which microbial activity then binds together. As the organisms glide and grow through sediment in order to re-establish themselves on the surface, they intertwine sediment particles and bind them into a coherent fabric.

Low flat mats are among the most efficient sediment traps. The pools they inhabit act as sedimentation basins; the sticky cyanobacterial surfaces retain even the finest particles. In cross-section, the low flat mats show alternating organic-rich and sediment-rich laminae. With every tidal cycle, a mat receives a load of sediment, and the microorganisms struggle to crawl on top of it until the next high tide. Those that do not make it are incorporated in the sediment as part of the organic-rich layer. Extracellular sheaths are invariably left behind. *Microcoleus chthonoplastes*, the principal constituent of such mats, is one of the fastest gliders.

There are many different heterotrophic bacteria in the anoxic zone beneath the mat that help degrade the organic matter produced above them. As long as oxygen-deficient conditions are maintained underneath the mat, the microbial processes there will necessarily be anaerobic. These include different types of fermentation, methanogenesis, and anaerobic respiration, mostly sulfate reduction. The gases that develop include hydrogen sulfide, methane, carbon dioxide, ammonia, and hydrogen. Although a complete anaerobic degradation of organic matter is possible, it involves several steps performed by different metabolic types. Since all necessary participants may not be in the right place at the right time, substantial organic residues are left to be buried and preserved. As a consequence, a relatively high rate of both inorganic and organic matter flux into the sediment beneath the low flat mat. This accelerates the filling of the pooled depressions. The mat is then elevated to a position where it will have longer exposure to air and desiccation.

A distinctive characteristic of intertidal microbial mats that are exposed to air is the presence of polygonal desiccation cracks. In smaller tidal channels, the cracking begins transverse to the flow; larger surfaces crack into polygons. The largest polygons form in the relatively moist center of channels, becoming progressively smaller toward the edges, where drying is faster (figure 6). As the points of contraction are evenly distributed over the sediment surface, the cracks develop roughly equidistant from these points; they then join to describe polygon shapes. The size of the polygons is determined by the severity of the dehydration and by the cohesiveness of the shrinking material. Continuing shrinkage causes upright curving of the polygon edges, leaving a central depression.

Water loss and shrinkage are common to the cracking of mud and microbial mats. The main differences in the cracking patterns of barren mud and mat-covered sediment originate in the differences in the cohesiveness of these materials. Mud cracks in straight lines, resulting in geometrically perfect polygons. These often subdivide into smaller secondary cracks. In mud cracks, each layer behaves independently relative to the position of the cracks. Cracks may fill with sediment but they cannot heal.

Figure 6
A desiccated tidal channel through the sabkha. Dried mats crack into characteristic desiccation polygons. Larger polygons form in the center of the channel; smaller polygons are found toward the edges.

Microbial mat, by contrast, behaves like a fabric of matted fibers, where the polygons form in ragged, uneven lines. Each polygon shrinks, but rarely does it develop a secondary crack. Microbial mats overgrow the edges of cracks and bridge them. In younger, weaker mats, the healed cracks are prone to cracking again in the same place, so that the cracks are perpetuated in subsequent layers. In vertical section, mat polygons can be seen stacked in prisms. Prism cracks have been recognized in the fossil record, and these described features are used to distinguish microbial mats from cracked mud.

Formation of the Fossil Record

The southern coast of the Persian Gulf is a *prograding* coast: the land is expanding, and the sea regressing, because sedimentary processes prevail over erosional ones. The geological terms *transgression* and *regression* refer to changes in the level of the sea in relation to the land. Transgression refers to the relative rise, and regression to the relative fall, of seawater level with respect to land. The 8000-year geological record at Abu Dhabi is rich in multiple cycles of alternating transgression and regression. Each cycle shows, in stratigraphic sedimentary sequence, a relatively rapid sea-level rise followed by a set of sediments that represent progressively shallower water.

A sedimentary sequence representing one such cycle can be observed in a core taken through the Abu Dhabi sabkha. These sediments have accumulated during the transformation of lagoon environments to sabkha plains. As the lagoons are gradually filled by sediments, new lagoons form offshore; the entire intertidal zone moves seaward, followed by the ever-expanding sabkha plain. Today's sabkha plain was once a lagoon; today's lagoon will someday become a sabkha plain. A core taken through the sabkha will reveal a sequence of sediments deposited on top of one another at different stages of this coastal progression. Using radiocarbon dating to trace this development, we learn that the process of sea-level decline lasted some 6000 years here, and that it was preceded by 2000 years of sea-level rise.

The level surface of the sabkha, initially created by the growth and sedimentation of microbial mats, is maintained by wind action and the capillarity of groundwater. When the groundwater level goes down, the surface soon dries out, crumbles, and is blown away by wind. When the groundwater level goes up, the surface is wetted by capillary action, and wind-blown sand grains stick to the sabkha surface. Thus capillarity keeps the sabkha surface at a certain distance from the groundwater level, maintained by alternating aeolian sediment input and erosion. The effect of this capillarity is to make the surface perfectly parallel to the groundwater level. The sabkha's groundwater is replenished from the land in the background by groundwater stored beneath the desert dunes, and by lagoonal floods brought by occasional storms. Because the groundwater slowly seeps toward the sea, its level and consequently the level of the sabkha are inclined slightly seaward. Because the fluctuation of the groundwater table is in the range of a few centimeters, so is the range of accretion and destruction of the sabkha surface. Anything buried below that fluctuating zone will not be affected, and the secrets of sabkha formation will be kept in the fossil record. Indeed, we know of a case of fantastic fossil preservation of comparable ancient microbial mats retained throughout the alpine orogeny. We now interpret the Lofer Formation high in the Austrian alps to be a fossil counterpart of the Persian Gulf sabkha sediments. We can infer that carbonate rocks in a cold mountain pass once were steaming microbial mat communities produced by *Microcoleus, Lyngbya* and other sabkha-making dwellers at the seashore.

Stromatolites of Shark Bay

What meanings, then, can we derive from our interpretation of the formation of microbial mats and stromatolites? We next travel halfway around the world to see an ancient living landscape, one that was very common in the Proterozoic past.

On the western coast of Australia the Indian Ocean enters a deep embayment called Shark Bay. There, a large hypersaline water body, Hamelin Pool, is separated from the outer bay by a shallow sandbar that restricts water exchange (figure 7). In the late 1950s two Australian geologists—a young professor, Philip Playford, and a graduate student, Brian Logan—explored Shark Bay. As they approached the shores of Hamelin Pool, a landscape of living stromatolites spread before them. The same sort of once-subtidal stromatolites were seen nearby at Carbla Point (figure 8). Formed by layers of microorganisms, the rocky shore was reminiscent of scientific reconstructions of Proterozoic life. The rocky structures were well known to these geologists but only as fossils. The record of life from the vast Archean and Proterozoic eons long before the appearance of any plant or animal life was replete with carbonate stromatolites.

Animals evolved about 600 million years ago. The Phanerozoic fossil record of animals and plants has been intensively studied by geologists and paleontologists for more than 200 years. The earlier span of life on Earth, between 3.5 billion and 600 million years ago, known now to be dominated by microorganisms, has only recently been subject to investigation. What did Playford and Logan see? Contemporary, live, growing stromatolites in the process of lithification!

I define stromatolites as rocks produced by the activities of microbes. How are living structures, composed of layered communities of microorganisms, turned to stone?

The fossil record of stromatolites is rich and diverse. Ancient stromatolites are abundant on the margins of Precambrian shields of all continents. They have been studied intensively in North America,

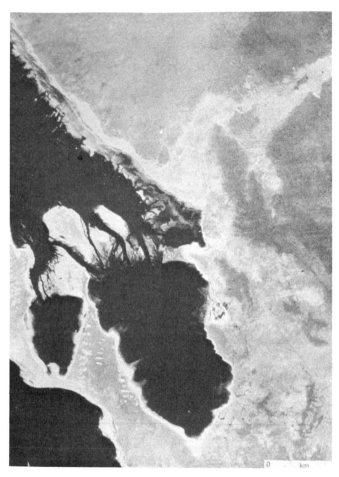

Figure 7
Shark Bay, a deep embayment on the west coast of Australia. This aerial photograph reveals the sandbars that separate Hamelin Pool, the larger body of water in the photo, from the rest of the bay.

Figure 8
The beach at Carbla Point. *Fragum* sand covers the beach between the stromatolitic heads.
Elso Barghoorn at left. *(Paul Strother)*

Africa, Asia, and Australia. The modern stromatolites discovered on
the shore of Hamelin Pool greatly resembled the ancient stromatolites
known to Playford and Logan from the Great Slave Lake area of
Canada. In that Arctic region, denuded by glaciers, the landscape is 1.8
billion years old, and the stromatolites can be observed in their original
positions.

A section through a stromatolite reveals alternating layers of organic-rich and mineral-rich material. The layers are interpreted to be
the cumulative record of the activities of microbial mats. We saw how,
in Abu Dhabi, these communities of organisms form mats by trapping
and binding sediment particles and sometimes precipitating minerals
that cement the particles. They can fortify the entire structure, in some
cases until stone is formed. These bacteria cope with rapid sedimentation by growing or crawling upward through each newly deposited
sediment layer to escape burial. The intertidal zone, where stromatolites commonly form, provides the conditions of moisture, nutrients,
and sunlight these microbes need.

Microbial mats are living communities of bacteria. A few form
stromatolites today. These are mainly in hypersaline basins all over the

Figure 9
Subtidal stromatolites in Exuma Sound, Lee Stocking Island, Bahamas.

world, including bays in western Australia; enclosed lagoons and ponds near Adelaide, in southern Australia; the coast of Kuwait, Saudi Arabia, and the United Arab Emirates; Solar Lake, on the Sinai Peninsula; bays along the coasts of the Red Sea; and lagoons in Baja California and on Christmas Island. Large subtidal ones form today off Lee Stocking Island in the Bahamas (figure 9). Stromatolites also form in many inland environments, including thermal springs in Yellowstone National Park, the Great Salt Lake of Utah, and the Cuatro Cienegas Basin in northeast Mexico. Saltworks in southern Europe and South America are also good environments in which to study modern stromatolites, although they do not necessarily form microbial mats.

Today's microbial mats dominate extreme environments, such as hypersaline ones. Mat-making microbes prefer to grow in conditions of normal seawater salinity. Why do they rarely form coherent mats or stromatolites there today? On the coasts of the Bahamas subtidal mats are destroyed by grazing and burrowing marine animals as quickly as they form. Stanley Awramik, a professor of geology at the University of California at Santa Barbara, compiled a record of the diversity of ancient stromatolites and found that their decline at the end of the Proterozoic eon coincides with the widespread appearance of mat-chomping marine animals in the fossil record. Mats persist today only under extreme conditions that preclude the presence of snails, worms, and other animals that feed on them.

Contemporary stromatolites have been discovered even in fresh water, such as the karstic springs of Cuatro Cienegas, Mexico. They also grow in normal marine environments such as Lee Stocking Island in Exuma Sound, the Bahamas. The absence of marine feeders cannot explain the growth and preservation of these stromatolites that share their habitat with a diverse, potentially threatening biota. The distribution of these stromatolites correlates with high rates of sediment cementation and lithification. I suspect that early hardening of the mat structures by mineral precipitates is the most likely key to their successful preservation.

Lithification takes place through intergranular precipitation of minerals such as calcium carbonate or silica. Calcification refers to lithification involving calcium carbonate; silicification refers to lithification involving silicon dioxide (figure 10). Other forms of lithification can involve oxides of metals such as iron and manganese. Bacteria and other microbial morphologies preserve their form better in silica than in the far more porous calcium carbonate rock. The main reason for this difference lies in the size of the crystal grains these minerals form. Calcium carbonate grains are generally much larger than those of silica, and larger than the microbes they enclose. Their size increases with every recrystallization. In the course of crystal growth, the microorganisms are displaced and crushed in the interfaces of colliding crystal planes. Silica crystallization generally starts with tiny opaline grains, minute cryptocrystalline hydrous opal, which embeds the microfossils without destroying them. This is particularly the case with fine fibrous crystals of the mineral chalcedony, which accounts for most of the cherty nodules within which preserved Archean and Proterozoic microbes are commonly found. Thus, microfossils have the best chance for preservation if the material in which they are embedded silicifies early, and if the surrounding rock retains its original hydrous mineralogy and does not recrystallize.

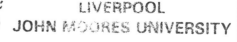

Figure 10
Stromatolites as lithified (carbonate) fossil microbial mats.

Entophysalis as a Living Fossil

The chances of long-term preservation and fossilization are better if mats escape lithification by carbonate precipitation and are silicified instead. Degradation in soft mats may be gradual and less destructive than in highly calcified structures. Characteristic morphological features of cell arrangements and colony construction are sometimes so well preserved as to remain recognizable even after hundreds of thousands of years of burial in sediment. Silicified Proterozoic microbial fossils well over a million years old were first discovered in the mid 1950s by Stanley Tyler and Elso Barghoorn. Since then, many such remains have been described, associated with fossil microbial mats recognizable in the rock record as stromatolites.

Remarkably, modern degraded *Entophysalis* (figure 11) from live mats does not differ much in shape and cell organization from the two billion year old Proterozoic microfossil *Eoentophysalis* (figure 12), which was found preserved in silicified stromatolites. The structures are seen in a petrographic thin section as clearly as if they were preserved in glass. The extracellular pigment, the multiple envelopes, and the granular remains of cells are all clearly visible.

Modern *Entophysalis major* and ancient *Eoentophysalis belcherensis* have many morphological features in common, some of which reveal similarities in their life cycles. The cell-division patterns revealed by the arrangements of extracellular envelopes are identical. Both organisms are stained darker outside and lighter inside by a similar light-responsive pigment. Both organisms formed carbonate stromatolites of approximately the same shape and size. Another indication of the similarity of their respective environments is the finding of pseudomorphs of halite (NaCl) within the ancient stromatolites. (*Pseudomorph* means "false form"; it refers to the result of a mineral replacement in which the crystalline form of the original mineral is retained.) Pseudomorphs of halite and gypsum are indicators of a hypersaline and/or evaporitic paleoenvironment. They tell us that during Proterozoic times the Belcher Islands were a part of a coast where stromatolites were forming in an intertidal setting similar to the coastal environments of today's Shark Bay.

Eoentophysalis has subsequently been found in younger Proterozoic strata in Amelia Dolomite in Australia (1.6 billion years old), the

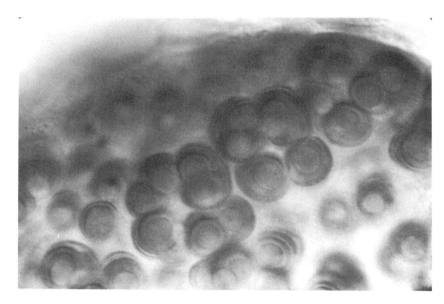

Figure 11
Light micrograph of living *Entophysalis* from a modern microbial mat.

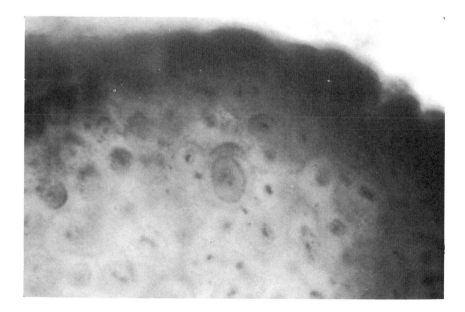

Figure 12
Petrographic thin section of *Eoentophysalis belcherensis* from a 1.8-billion–1.9-billion-year-old silicified stromatolite from Belcher Island, Hudson Bay, Canada. Paired cells in upper left corner and central region are evidence of cell division.

Gaoyuzhuang Formation in China (1.2 billion years old), the Bitter Springs Formation in Central Australia (800—900 million years old), and in Greenland (700 million years old). Today *Entophysalis* inhabits intertidal zones in the bays of subtropical coasts, such as Shark Bay or the lagoons of the Arabian/Persian Gulf. Its occurrence during the Proterozoic eon encompasses a time span much longer than that which has lapsed between the beginning of the Phanerozoic and the present. *Entophysalis,* the oldest known living fossil, has maintained its form and its function through a geological time span exceeding half the entire history of life on Earth. Living microbial mats and their lithified remains, stromatolites, are among the most valuable clues we have for our reconstruction of planetary conditions during the vast span of history before the origin and evolution of any plant or animal.

Readings

Cohen, Y., R. W. Castenholtz, and H. O. Halvorson, eds. 1984. *Microbial Mats: Stromatolites.* Alan R. Liss.

Cohen, Y., and E. Rosenberg. 1989. *Microbial Mats: Physiological Ecology of Benthic Microbial Communities.* American Society for Microbiology.

Fisher, A. G. 1964. The Lofer cyclotherm of the Alpine Triassic. *Kansas Geologic Survey Bulletin* 169: 107–149.

Golubic, S. 1976. Organisms that build stromatolites. In *Stromatolites* (Developments in Sedimentology, volume 20), ed. M. Walter. Elsevier.

Golubic, S. 1976. Taxonomy of extant stromatolite building cyanophytes. In *Stromatolites* (Developments in Sedimentology, volume 20), ed. M. Walter. Elsevier.

Golubic, S. 1985. Microbial mats and modern stromatolites in Shark Bay, Western Australia. In *Planetary Ecology,* ed. D. Caldwell et al. Van Nostrand Reinhold.

Golubic, S., and H. J. Hoffmann. 1976. Comparison of modern and mid-Precambrian Entophysalidaceae (Cyanophyta) in stromatolitic algal mats: Cell division and degradation. *Journal of Paleontology* 50: 1074–1082.

Lowenstam, H. A., and S. Weiner. 1989. *On Biomineralization.* Oxford University Press.

Walter, M. R. 1976. *Developments in Sedimentology 20: Stromatolites.* Elsevier.

Westbroek, P., and E. W. De Jong, eds.. 1983. *Biomineralization and Biological Metal Accumulation.* Reidel.

8 Symbiosis and the Origin of Protists

Lynn Margulis

Here Lynn Margulis describes her current view of the origin of eukaryotes (organisms composed of cells with nuclei). The earliest eukaryotes were anaerobic protists; a few have living but reclusive
descendants in today's world. Some of the protist dwellers in the
anoxic world, Margulis believes, were ancestors of plants, animals, and
fungi. All organisms with membrane-bounded nuclei are thought
to have evolved from the genetically integrated bacterial symbionts
that became the very first swimming anaerobic protists. Some of these
later acquired mitochondria and plastids from, respectively, oxygen-
respiring and photosynthetic bacterial symbionts.

The serial endosymbiosis theory (SET) of the origin of cells with nuclei
has engaged me all of my professional life. Although not entirely
proven, most of its postulates are widely accepted. Nucleated cells,
those constituting animals, plants, fungi, and protoctists, are, I argue,
derived from tightly integrated bacterial communities. Eukaryotes did
evolve from bacteria, but not directly. A single eukaryotic cell differs
from a single bacterial cell (figure 1).

Bacteria may be small, but they are just as "high" as any other life
form. Bacteria evolved before any eukaryotes. Extant survivors of
more than 3 billion years of evolution, they have an astounding range
of metabolic diversity—far greater than that of animals and plants.
Bacteria cycle all of the chemical elements required by life as biogenic
gases and solutes. They tolerate extremes of oxygen, temperature, and
pressure better than the most resistant eukaryotes. Prokaryotic (that is,
bacterial) cells are units: membrane-bounded, protein-synthesizing,
self-sufficient little systems. The cells of all bacteria are prokaryotic. All
have DNA and various types of RNA, including messenger, ribosomal,

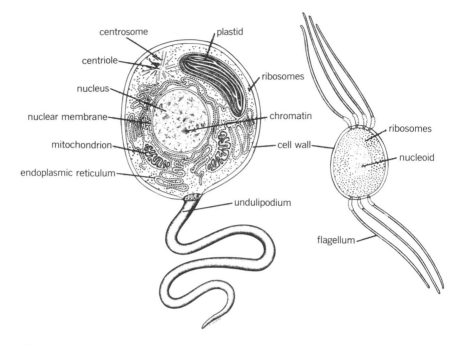

Figure 1
Eukaryotic and prokaryotic cells compared. Eukaryotic cells (left) contain membrane-bounded nuclei and other organelles, including mitochondria, plastids, and undulipodia (complex motility organelles composed of tubulin and many other proteins). Prokaryotes (right) have DNA that is not surrounded by a membrane, and flagella with shafts composed of a single protein (flagellin). *(Christie Lyons)*

and transfer RNA. All possess ribosomes and are bounded by a plasma membrane consisting of lipid and protein. A minimal cell, with at least 500 genes and their products, constitutes the unit bacterial cell. Cell walls, if present, always lie outside the plasma membrane. With respect to their limiting structures, bacteria can be grouped into four classes: Gram-negative, Gram-positive, Gram-variable, or without walls at all.

Organelles from Free-Living Bacteria

The SET states that the three classes of organelles of eukaryotic cells—undulipodia (cilia and other microtubular organelles of motility), mitochondria (organelles of energy transduction), and plastids (organelles of photosynthesis)—originated from bacterial symbionts. If this view is correct, animal cells all have at least three kinds of bacterial ancestors, and plant cells have at least four. All eukaryotes are chimeras.

The phylogeny or "family tree" of eukaryotic cells is depicted in figure 2. The monerans (all the bacteria) are shown at the bottom. We move chronologically from past to present as we ascend the paths of this diagram. Toward the top are the four groups of eukaryotes: plants, animals, fungi, and the ancestral protoctists.

Cells evolved by "serial symbiosis," a term that refers to the acquisition and integration of particular bacteria in a definite sequence. Symbiosis means the long-term physical association of organisms of different kinds. In the SET described here, the processes that established eukaryotic cells were symbioses that led to symbiogenesis as the symbiotic associations became permanent. Two or more types of symbionts coevolve with time and become genetically integrated to form a single new organismal unit. This, in turn, is capable of acquiring still other symbionts.

I hypothesize that the first step in the origin of the eukaryotes—the process of eukaryosis—was the permanent fusion of an archaebacterium much like the *Thermoplasma* of today with a motile eubacterium much like *Spirochaeta* of today. This first crucial step that led to the origins of the nuclear membrane system and of intracellular motility is detailed below. The *Spirochete*-plus-*Thermoplasma* integration, I believe, led to a type of protist with modern descendants. These were "archaeprotists," namely mastigotes—swimming cells that lack mitochondria. Some of the descendants of these protists acquired oxygen-respiring bacteria, the symbionts that became the mitochondria. Whereas fungi and animals evolved directly from aerobic mastigotes, the organisms that became algae and plants acquired further symbionts: cyanobacteria.

The Origin of Nucleocytoplasm

The term *nucleocytoplasm* refers to the single, integrated system of the eukaryotes, which is composed of the membrane-bounded nuclear DNA and the ribosome-rich cytoplasm external to the nucleus. Sequences of nucleotide base pairs in nuclear DNA determine complementary sequences in messenger RNA. In short lengths, mRNA travels

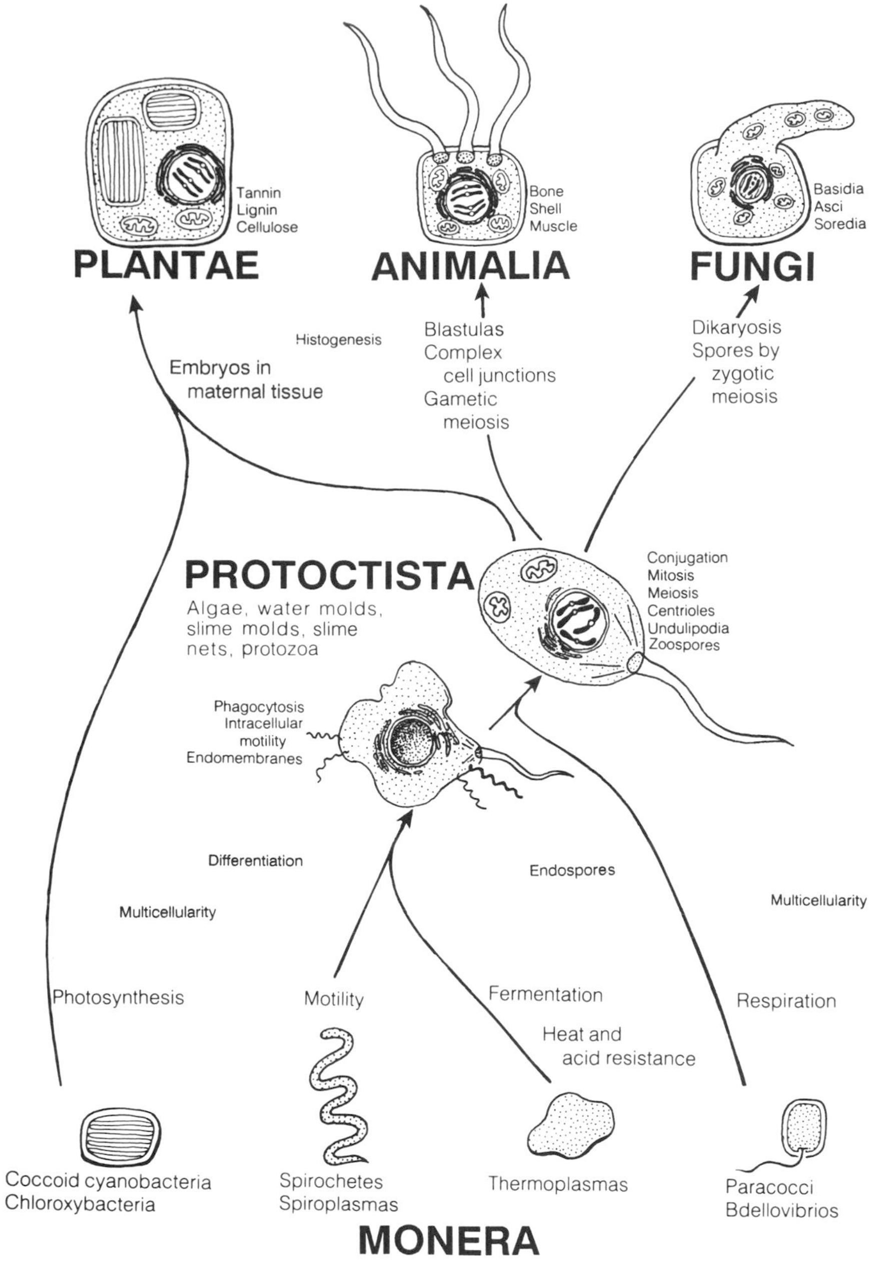

Figure 2
Cells as microbial communities: the serial endosymbiosis theory of the origin of eukaryotic cells. *(Kathryn Deslisle)*

to the cytoplasm, where the messages are "read" on cytoplasmic ribosomes in such a way that specific proteins are synthesized. No eukaryotic cell lacks this nucleocytoplasmic system. Thus, a central question in eukaryosis is "What is the evolutionary ancestry of the nucleocytoplasm?"

A veritable revolution, initiated by Carl Woese of the University of Illinois and his colleagues, has occurred in the identification of relationships not only among live bacteria but also between bacteria and nucleated organisms. Today hundreds of scientists compare genes (DNA) to learn about the microbes from which DNA molecules were extracted. Through comparison of the specific gene that determines a universally distributed ribonucleic acid molecule (16S ribosomal RNA), certain relationships between live organisms can be elegantly assessed and inferences about their ancestry can be made. From an enormous data base of genes for ribosomal RNA taken from thousands of bacteria and eukaryotes, Woese has asserted that only three evolutionary lineages of life (he calls them "Domains") led to all extant forms. Those three include the two groups of prokaryotes ("Archea" microorganisms like methanogens, thermoacidophiles, and halophilic bacteria) and nearly all the other groups of bacteria (cyanobacteria and other photosynthesizers, enterobacteria like *Escherichia coli*, myxobacteria, spirochetes, and so forth), and, in Woese's third group, all the eukaryotes (plants, animals, protoctists, and fungi). In this analysis, the three ancient lineages (Archea, for which I retain Woese's earlier name "Archaebacteria" because in every way these microbes are bacteria) include "Eubacteria" (the term used by Woese for all the bacteria not classified as Archaebacteria), and "Eukarya" (the group comprising all the nucleated organisms). Many scientists concur with the Woesian three-domain scheme, but I reject it. Although I entirely applaud the contribution of molecular comparisons to clarification of relationships between groups of microbes, to me and to Karlene Schwartz both "Archaebacteria" and "Eubacteria" are inside the great group, the kingdom "Monera" (which is the same as "Bacteria" or "Prokaryotae"). We recognize archaebacteria and eubacteria as two subkingdoms of the great early divergence of bacteria. The eukaryotes, all of them, differ in principle. Eukaryotes, all of them, evolved symbiogenetically. They are products of fusion of at least two kinds of bacteria. Indeed, most eukaryotic cells evolved from fusion between more than two kinds of bacteria. In my opinion, the first fusion, one between an archaebacterium and a eubacterium, led to the differentiation of the nucleus from the cytoplasm. Symbiogenesis among bacteria was the most important process in the evolution of the "Eukaryota." Since the appearance of the membrane-bounded nucleus (the structure that defines eukaryotes) accompanied the origin of the entire group based on eukaryotic cells (protoctists, animals, plants, and fungi), only two domains of life deserve recognition: the prokaryotes (archaebacteria and eubacteria, non of which evolved symbiogenetically) and the eukaryotes (or "Eukarya"), all of which are products of thoroughly integrated symbioses.

The leading hypothesis, at least at the University of Massachusetts, in spite of Woese's great contribution, is that the nucleocytoplasm originated symbiogenetically from eubacteria and archaebacteria like the *Thermoplasma acidophilum* cultured in the laboratory of Dennis Searcy at the University of Massachusetts in Amherst (figure 3). *Thermoplasma acidophilum*, the best-studied of the thermoplasmas, belongs to the Archaebacteria. The possibility that a *Thermoplasma*-like bacterium evolved into nucleocytoplasm is now being tested by Searcy. Why might *Thermoplasma*-like bacteria be related to the ancestral nucleocytoplasm? *Thermoplasma* apparently contains histone-like proteins. Nearly absent from bacteria, histone proteins are distinctive features of eukaryotes. Other features of eukaryotes thought to have evolved from *Thermoplasma* are tabulated in figure 4.

Although irregular in shape, *Thermoplasma*, like all other bacteria, has no nucleus and does not divide by mitosis. If we accept Searcy's idea that *Thermoplasma*-like microbes became nucleocytoplasm when they established early symbioses with spirochetes, our next question involves the origin of the nucleus itself. Minimally, a nucleus must be bounded by a nuclear membrane. I suspect that the nuclear membrane evolved as a consequence of symbiotic associations between *Thermoplasma*-like bacteria and spirochetes that penetrated them. The parrying led to a proliferation of membrane, including endoplasmic reticulum, inside what had been *Thermoplasma*. Newly formed nuclear membrane became involved in the segregation of *Thermoplasma* DNA in the way that membrane originally segregated DNA in *Thermoplasma* before it was beset by would-be symbionts.

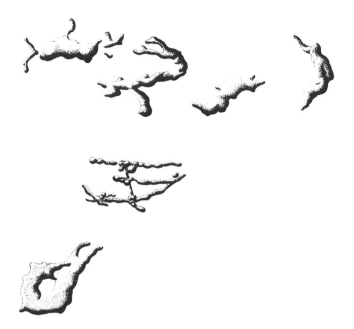

Figure 3
Thermoplasma acidophilum, drawn from electron micrographs. *(Christie Lyons)*

The nuclear membrane is just one among many membrane elaborations characteristic of eukaryotic cells. It is continuous with the other cell membranes: endoplasmic reticular membrane, Golgi-apparatus membrane, the outer membrane of the mitochondria, and the plasma membrane. Membranes are dynamic—they should be thought of as moving belts of fusing and rupturing lipid layers in which proteins are embedded and ions are pumped. Membrane is continually produced in the nucleus at DNA attachments and is moved outward; unlike the static drawings typical in textbooks, membranes are fluid, interactive, and asymmetrical. Since in most organisms membranes tend to grow vigorously at the site of microbial invasion, I suspect that intracellular membranes in general proliferated in response to once-aggressive microbial associations.

At the beginning of any symbiotic associations, the genomes of the unit cells are separate. Host and invading bacterial symbiont each have their own genomes. Integration of genomes probably occurs only after a great deal of other structural integration. Integration of the genomes

Spirochaeta   *Thermoplasma*

Spirochaeta	Thermoplasma
motility	acid, heat resistance
centriole/kinetosome	ATP synthetase, PGkinase
microtubules	chromatin
eubacterial genes, proteins	archaebacterial genes, proteins
hsp70, axonemes	EF2, EFTu, lamins
O_2 hypersensitivity	O_2 microaerophilia
retraction	pleiomorphism
H_2 production	sulfur reduction

Figure 4
Eubacterial (spirochete) and archaebacterial (thermoplasm) contributions to the origin of nucleocytoplasm (by hypothesis).

of former symbionts is probably a consequence of moving "small replicons." Plasmids, viruses, transposons, and DNA in solution are all examples of small replicons—small lengths of replicating DNA that can be manipulated by enzymes. Such small DNA replicons can move and be incorporated into "large replicons," longer pieces of DNA. Since mechanisms that incorporate plasmids or viruses into larger pieces of DNA are ubiquitous in bacteria, I hypothesize that integration of small replicons from bacterial symbionts into what became nuclear DNA was an essential aspect of eukaryotic origins.

With the acquisition of motile spirochete bacteria by a *Thermoplasma*-like bacterium, the first step toward eukaryosis was taken. Descendants of such a first step toward symbiosis still thrive in anaerobic muds and as gut symbionts in insects. Many extant mastigotes have microtubule-composed undulipodia and their kinetosomes still connected to the nucleus. In all archaeprotist taxa they all lack mitochondria. Therefore, the most tenable view is that the thermoplasma-spirochaeta fusions become nucleocytoplasm—undulipodia before mitochondria were acquired. Hypermastigotes (including *Staurojoenina*, shown in figure 5), devescovinids, and mastigotes are examples of

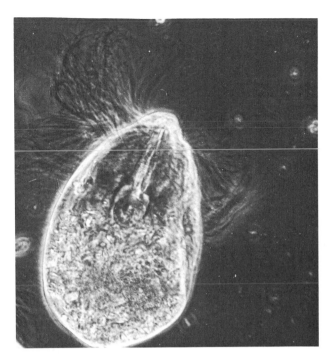

Figure 5
Staurojoenina, a hypermastigote protist symbiotic in termite hindguts, possesses undulipodia and lacks mitochondria. *(Christie Lyons)*

these; they lack mitochondria yet have undulipodia. The kinetosome
centriole microtubule systems, nearly universal in eukaryotes, are present in all mastigotes. From the mastigotes (undulipodiated organisms
at the protoctist level of organization) to the origin of fungi, animals,
and plants, an uncountable series of evolutionary steps occurred, leading to many innovations including the origins of standard animal-plant
mitosis, sperm tails and meiosis.

Mitochondrial Origins

The next step in this saga was the origin of mitochondria. Whereas the
symbiotic acquisition of Gram-negative eubacteria as spirochetes is
hypothesized to be central to the origin of mitosis and all other forms
of intracellular motility, the acquisition of mitochondria led to the
colonization of the aerobic world by eukaryotes. The earliest mastigotes became aerobes when they either engulfed or were attacked by

oxygen-respiring, eubacteria that became the organelles called mitochondria. The bacteria that evolved into mitochondria probably resembled members of today's genera *Paracoccus, Bdellovibrio,* and *Daptobacter.* Such acquisition of symbiotic, intracellular, oxygen-respiring bacteria probably occurred many times; at least one lineage was spectacularly successful.

All plants, algae, fungi, and animals—more than 30 million species—contain mitochondria in each of their cells. This observation suggests that the common ancestor of animals, plants, fungi, and algae—presumably some aerobic protist—possessed mitochondria. Mitochondria, universal in algae and all other photosynthetic eukaryotes, were already present in cells when plastids were acquired. The original eukaryotes were heterotrophs unable to make their own food. They required preformed organic compounds, or other whole organisms, as food and energy sources. Some of the early protists phagocytotically ingested, but did not digest, photosynthetic bacteria. Photosynthetic bacteria—coccoid cyanobacteria with chlorophyll b pigment such as *Prochloron,* and other phototrophs—began as food for translucent protists. They become plastids. One of the last steps in the origin of algal and plant eukaryotes was the acquisition by symbiosis of photosynthesis.

The acquisition of photosynthetic symbionts was the most recent step in my version of the serial endosymbiosis theory, because the other two classes of organelles—undulipodia and mitochondria—are thoroughly integrated in all cells that contain plastids. This last evolutionary step, the acquisition of plastids, is the easiest to document since the plastids retain many features of free-living cyanobacteria.

A remarkable aspect of this phylogeny—which by 1988 had reached its current form, with undulipodia preceding mitochondria (figure 2)—is that it was anticipated in its fundamentals by K. S. Merezchkovsky nearly 100 years ago. The phylogeny of all life forms drawn by this Russian biologist (a professor at the University of Kazan at the time) is depicted in figure 6. Merezchkovsky placed the origin of the bacterial groups at the far left. Following the "tree of life" drawn by the German scientist Ernst Haeckel in 1865, he called the microorganisms *monerans.* From the set of symbioses depicted as broken lines running from lower left to right, it is apparent that Merezchkovsky thought that large cells had emerged from symbiotic associations. He

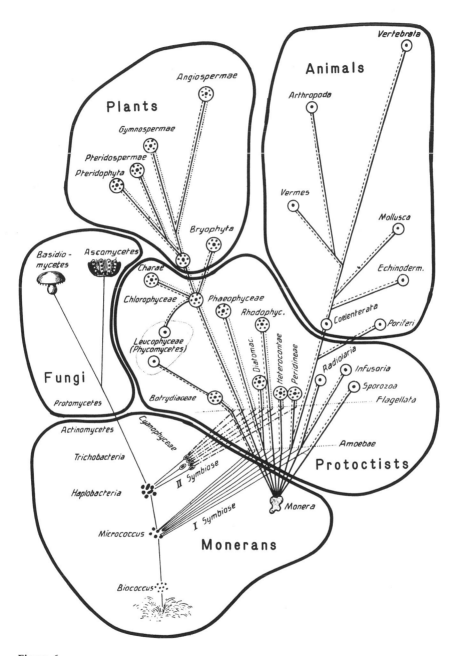

Figure 6
Merezchkovsky's anastomosing phylogeny based on symbiosis. The current five-kingdom classification scheme is superimposed.

derived all "infusoria" and other protists, as well as all animals, from double symbioses. ("Infusoria" is an old name for ciliates.) He showed algae and plants evolving from triple symbioses (nuclei, mitochondria, and plastids, in that order). As far as I know, this fundamental conceptualization of symbiosis, on which I have worked all my professional life, was first developed by Merezchkovsky before 1910. Merezchkovsky's contribution is little known in the West. Similar ideas of symbiogenesis—the origin of evolutionary novelty through symbiotic alliances—were put forward in the United States by Ivan E. Wallin, a professor of anatomy at the University of Colorado, who wrote a fascinating small book called *Symbionticism and the Origin of the Species* in 1927. Thoroughly rejected in its time, Wallin's work remains poorly known. The history of these symbiogeneticists is detailed in L. N. Khakhina's 1992 book *Concepts of Symbiogenesis*.

Control of Symbionts by Hosts

In symbioses the nature of the relationships change. Certain bacterial associations begin as lethal pathogenic infections. They end under control of the host cells. *Amoeba proteus*, grown by Kwang Jeon at the University of Tennessee, is an example of this. The amebas in Jeon's laboratory became infected by bacteria; the consequence was that nearly all the amebas died. Electron-microscope investigation showed that tiny black specks within the infected amebas were bacteria. At first the huge numbers of bacteria inside the amebas' cytoplasm killed nearly all the amebas; however, later the surviving bacteria became symbionts.

How did the pathogenic bacteria kill the amebas at first? This is not known in detail, yet Jeon saw that the amebas fed by taking bacteria into their phagocytotic vacuoles. He showed that digestive enzymes, which kill bacteria, are excreted across the vacuolar membranes. Though very few can, some bacteria debilitate or attack amebas in a way that completely precludes the efficacy of the amebas' digestive enzymes. Such bacteria, since they are not digested, actually penetrate the vacuole membranes and enter the cytoplasm. They find themselves "sitting in the pantry" and start growing. In the early days of Jeon's discovery, great populations of intracellular resistant bacteria were found in ameba cytoplasm—more than 120,000 bacteria per ameba. If the bacteria divide again, before the ameba does, the number of bacteria per ameba goes from 120,000 to 240,000 and the ameba dies. An amazing corollary—which I have only imbibed by hearsay—is that this microbial behavior is extremely similar to that of the bacteria that caused "Legionnaires' disease." Apparently, then-unknown pathogenic bacteria entered the cooling water of the air-conditioning system of a Philadelphia hotel in which an American Legion convention was taking place. Once transmitted to a human, the bacteria were engulfed by white blood cells of the immune system. A number of Legionnaires became fatally ill when their white blood cells were killed by pathogenic bacteria. Free-living amebas are similar in morphology to the lymphocytes in human blood. *Legionella* bacteria penetrate the vacuoles of our ameboid lymphocytes, the cells our bodies use to restrain bacteria. *Legionella* generally grows in the cytoplasm more rapidly than the lymphocytes can contain it. Lymphocytes, like Jeon's amebas, are under threat of death—they endure enormous selection pressure as they produce inhibitors, shut down DNA and protein synthesis, and behave in other ways to restrain intracellular growth of pathogenic bacteria in themselves. In most cases, the infections continue and the amebas or ameboid lymphocytes are destroyed by bacteria growing inside them.

However, bacterial growth can be restrained by the infected cells. Jeon isolated some amebas in which the quantities of intracellular pathogenic bacteria dropped from 150,000 bacteria per ameba to fewer than 40,000. The infection, in these cases, concomitantly became benign. After 5 years many live amebas contained only 40,000 bacteria per cell; Jeon showed that the descendants of the amebas that had survived the bacterial infection now required those same bacteria for their health and growth. What had once been a pathogenic bacterial infection had—by definition—become a symbiosis; moreover, the formerly invasive bacteria were now organelles! The nucleocytoplasm of the long-infected amebas no longer could survive without the bacteria. This symbiosis was established in only a few years; certainly such events can take place over geological periods of time. Jeon's story is instructively analogous to the origin of mitochondria: pathogenic free-living bacteria infect other cells and become cytoplasmic resident bacteria under control of these cells. A first step in the origin of any obligate endosymbiosis is the cell's restraint of the intracellular symbiont's tendency to continue to grow. In the final stages in the development of these relationships, symbiotic assimilation occurs such that the bacteria unequivocally become organelles required for the continued growth of the host cell.

Photosynthetic Animals

Many examples exist in natural history where a heterotrophic organism eats but fails to digest a photosynthetic organism and the two integrated evolve as a consortium. *Hydra viridis,* found in fresh waters all over the world, provide excellent illustrations of heterotrophic organisms that ate and failed to digest algae. The endodermal (inner) layer of the hydra in nature contains the coccoid green algae *Chlorella.* The same animal, *Hydra viridis,* can be treated with chemicals and high-intensity light to induce loss of its *Chlorella.* An experimentally derived white hydra, unfed and placed in the dark, will die within days. An unfed green hydra in the dark dies in a few days too. A white hydra, experimentally deprived of its *Chlorella* and placed in the light without any food (*Artemia*—brine shrimp—are their usual food) will also die in a few days. However, green hydra under starvation conditions continue to live in the light for at least 3 or 4 months. If given *Chlorella* as food, a white hydra will "regreen"—the *Chlorella* it eats will not be digested but will become symbionts and will continue to help feed the hydra as long as light is available. Such associations between photosynthetic and heterotrophic entities serve as models for the origin of plastids in the algal ancestors of plants.

More controversial, and still largely unaccepted relative to the origin of plastids and mitochondria, is the idea of the origin of undulipodia (cilia) from spirochetes. The detailed structural similarity of the kinetosomes and axonemes that comprise undulipodia whether of cilia or sperm tails suggests that this motility structure evolved only once. We simply do not yet know in detail how the complex motility of the eukaryotic cell originated.

Symbiogenesis Theory

Why was Merezchkovsky's theory of symbiosis never accepted? Why is Wallin's theory of the origin of species by symbiosis not generally known? The complex reasons were analyzed by Jan Sapp (1987). Bacteria are associated with disease and food contamination, not with the generation of evolutionary novelty. However, the concept of "symbiogenesis" and its evolutionary importance is entering mainstream biology. In another decade or so the idea that all eukaryotic cells are genetically integrated, tightly coevolved communities of bacteria will, I predict, be fully accepted by all serious scientists.

Undulipodia and Flagella Compared

The most difficult and controversial aspect of the symbiotic theory of the origin of eukaryotic cells is the hypothesis that undulipodia evolved from spirochetes in the first merger that led to protists. However, this proposition is intrinsic to the theory.

Undulipodium is the generic name for the [9(2) + 2] microtubular structure, the motility organelle common to many eukaryotic cells and entirely absent from all prokaryotes (figure 7). *Undulipodia*, which differ from bacterial flagella (figure 5) in all respects, have many names: cilia, flagella, sperm tails, and even pecilokont. Knowledge of undulipodia is crucial to an understanding of the origin of mitosis, the cell-division process unique to eukaryotes and the subsequent evolution of meiotic sexuality.

All undulipodia are composed of nine doublets of microtubules, the walls of which are made of α and β tubulin proteins. In amino acid composition, these tubulins are extraordinarily similar to the α and β tubulins found in the mitotic spindles of animals, plants, and fungi. Each undulipodium is always underlain by its kinetosome, the [9(3) + 0] microtubular structure. The hypothesis is that the entire undulipodium, including its kinetosome, originated as a bacterium and its attachment structure. Before evolving into a cell organelle, the undulating microbe was free-living and of course highly motile. The undulipodium originated as a free-living microbe similar to many present-day spirochetes that attach to other microbes. Eventually the spirochetes and their would-be victims integrated. Some nucleic acids, certain motility and other proteins were redeployed as the partners merged and new sorts of larger, more complex swimming cells—anaerobic mastigotes—evolved. A series of selective steps ensued that led to the evolution of many different species of eukaryotic microbes. Such protists were ancestors of animals, plants, and fungi and of their meiotic sexuality.

In the course of evolution, constituents of the attached motile bacteria were used for many purposes. Membranes fused; some were lost.

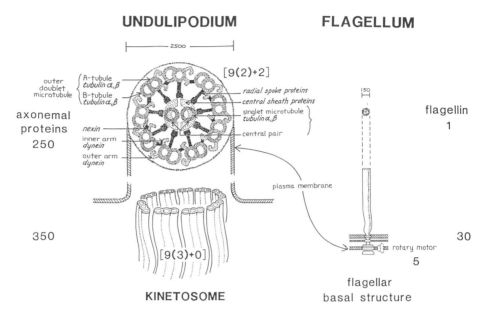

Figure 7
Undulipodium and flagellum compared. The undulipodia of eukaryotic cells are relatively large, complex structures. Approximately 250 nanometers in diameter, they are composed of some 600 proteins: 250 in the axoneme and 350 in the underlying kinetosome. Bacterial flagella, only 15 nm in diameter, contain only one protein (flagellin) in their shafts. The rotary motor is made up of about five proteins involved directly in motility and 25 flagella-assembly proteins; a total of 30 proteins are found in the whole flagellar basal structure.

The motile proteins moved around in the *Thermoplasma*-like bacteria that became mastigotes. Eventually microtubules, which had originally entered as part of the motility apparatus of the spirochetes, were used in the process of mitosis; they became mitotic microtubules.

Microtubule Structures: Spindles, Kinetosomes, Centrioles, and Undulipodia

The scenario by which free-living motile bacteria were transformed into mitotic structures (figure 8) inside what became the protist cell is detailed at the professional level in my book *Symbiosis in Cell Evolution* (second edition, 1993). The story is also told in chapter 4 of *Microcosmos* and in chapter 6 of *What Is Life?*

The search goes on. Although 30 years of investigation have not established the SET fully, no other theory has rivaled it in comprehen-

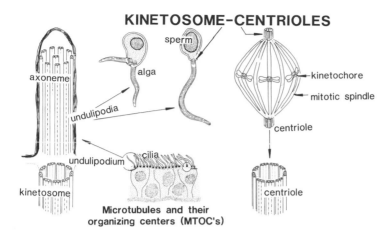

Figure 8

Relation of kinetosomes, centrioles and spindles, [9(2) + 2] microtubules structures in undulipodia, sperm tails, cilia, and ciliated epithelium. Kinetosomes and centrioles are homologous; both have a [9(3) + 0)] structure. *(Kathryn Deslisle)*

siveness and predictive power. Whatever the details, the concept of the eukaryotic cell as a consortium, a chimera, an integrated system of composites, seems to be as firmly established within biological science as the theory of plate tectonics is within geological science.

Readings

Gray, M. W. 1985. The bacterial ancestry of plastids and mitochondria. *Bio-Science* 33: 693–699.

Khakhina, L. N. 1992. *Concepts of Symbiogenesis.* Yale University Press.

Margulis, L. 1993. *Symbiosis in Cell Evolution,* second edition. Freeman.

Margulis, L., and D. Sagan. 1995. *What Is Life?* Simon & Schuster.

Margulis, L., and D. Sagan. 1997. *What Is Sex?* Simon & Schuster.

Margulis, L., and Sagan, D. 1997. *Slanted Truths: Essays on Gaia, Symbiosis and Evolution.* Springer-Verlag.

Margulis, L. 1998. *Symbiotic Planet.* Basic Books.

Sapp, J. 1987. *Beyond the Gene.* Cambridge University Press.

Woese, C. R., O. Kandler, and M.L. Wheelis. 1990. Towards a natural system of organisms: Proposal for the domains Archaea, Bacteria and Eucarya. *Proceedings of the National Academy of Sciences* 87: 4576–4579.

9

The Antiquity of Life: From Life's Origin to the End of the Lipalian Period

Mark McMenamin

Approximately 600 million years ago a marine life form suddenly appeared upon Earth without any trace of prior evolution. Defying any normal classification, these Ediacarans were similar to animals but never passed through an embryonic stage. The mystery of what the Ediacarans are and how they evolved is explored here by Mark McMenamin, a professor of geology at Mount Holyoke College, who has annotated and made current the definitive English-language edition of Vladimir I. Vernadsky's great book *The Biosphere*. Using novel ideas about the late pre-Phanerozoic supercontinent Rodinia, McMenamin discusses the effects of that paleoenvironment on the subsequent evolution of large life forms.

Contemplating the origin of life requires a confrontation with an alien planet, namely Earth before Biosphere. The great Russian geochemist Vladimir Vernadsky was acutely aware of this problem. He maintained that life is eternal (at least insofar as we may contemplate our planet in its present state, with its pervasively bio-influenced crust). In other words, for Vernadsky life is such a potent geological force that, in a nontrivial sense, before life existed Earth did not exist.

In the first edition of this book, Elso Barghoorn argued that "the study of the origin of life as an event is more a philosophical than a scientific pursuit." Clearly, however, the geological sciences can provide constraints on when the event must have occurred on Earth's surface (or alternatively, when life must have been delivered here from elsewhere). Life evolved or arrived on this planet sometime between 4.0 billion and 3.5 billion years ago.

The most ancient evidence for life, discovered by Stanley A. Awramik of the University of California at Santa Barbara, occurs in a sequence of Archean rocks of Western Australia known as the Warra-woona Group. Dated to 3.5 billion years of age, these fossils consist of cellularly preserved filamentous microbes. The fossils have been given the names *Archaeotrichion septatum*, *Eoleptonema australicum*, *Primaevifilum septatum*, and *Archaeosillatoriopsis disciformis*, but these long-winded monikers mask the fact that, at least in terms of their cellular morphology, these most ancient of fossils are virtually indistinguishable from living cyanobacteria, also known as pond scum.

Pond scum bacteria, and the microbial mats formed by their densely settled communities, are the primary players in the biotic scene in the several billion years after the appearance of life on this planet. Fossils of these bacteria are rare during this interval, but when found they are usually associated with stromatolites (figures 1 and 2). Stromatolites are laminated organosedimentary structures thought in many cases to have been formed by the sediment trapping and carbonate secreting abilities of the microbial mat bacteria.

Archean (3.8 billion to 2.5 billion year old) stromatolites occur in sediments deposited within the photic zone, and it has been reasonably inferred that the bacterial tenants of these stromatolites were photosynthetic. The biochemistry of living cyanobacteria, some of which greatly resemble the Archean fossil bacteria, has a distinctive signature indicative of a first appearance in a hot environment. Chlorophyll is apparently descended from an ancient heat-sensing biomolecule. Ribulose-1,5-biphosphate (also called Rubisco or RuBP), the molecule that catalyzed the linking of carbon to other atoms (carboxylation) and hence a chemical sine qua non of life, requires for its biosynthesis chaperonin. Chaperonin was descended from heat shock proteins. Thus, the Archean stromatolite-forming communities represented, early in life's history, an expansion of life from marine hydrothermal vent habitats (many of which presumably had low ambient levels of sunlight) into fully illuminated shallow marine habitats. This is very much in accord with the expansive properties of living communities that Vernadsky called "the pressure of life," and it presaged a later expansion of life up the gradient of the hypsographic curve—the development of Hypersea (eukaryotic land ecosystems) approximately 400 million years ago.

At the end of the Archean eon, the Earth's continents reached a critical size that allowed the formation (by sagging of the edge of each

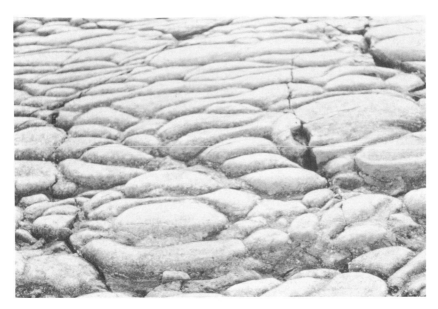

Figure 1
Ancient stromatolites (2 billion years old) at Great Slave Lake, Northwest Territories, Canada. *(From S. Golubic, "Microbial mats and modern stromatolites in Shark Bay, Western Australia," in* Planetary Ecology, *ed. D. Caldwell et al. (Van Nostrand Reinhold, 1985). Copyright 1985 Van Nostrand Reinhold. Reprinted with permission.)*

Figure 2
Recently formed stromatolites at Carbla Point, Shark Bay, Western Australia. *(Paul Strother)*

continental edge in such a way that part of the edge is covered by a relatively shallow layer of seawater) of extensive continental platforms. Stromatolitic reefs and banks flourished in this new habitat. The stromatolites and their microbial formers, no longer restricted to relatively narrow coastal strips of appropriate water depth, began to aid in the deposition of thick Proterozoic sheets of microbially precipitated calcium and magnesium carbonate sediments. Individual stromatolites the size of small mountains began to form in these environments after the end of the Archean eon (2.5 billion years ago).

The first great change we perceive in the fossil record occurred shortly after the globally distributed marine deposition of banded iron formations (peak deposition approximately 2 billion years ago) and the oxygenation of the atmosphere. This oxygenation event, which occurred as a result of the increasing abundance of oxygen-generating stromatolites, is called the "Oxygen Crisis," really an anthromorphism of the "feelings" presumed to have been experienced by the anaerobic bacteria living at the time. The crisis ended when remaining anaerobes were driven to low-oxygen environments and the first fully aerobic organisms (protected by enzymes such as superoxide dismutases with strong affinities for oxygen radicals) appeared to repopulate the environments now saturated with oxygen. "Necessity is the mother of invention" was never truer than at this point in Earth history, unless of course the superoxide dismutates had already evolved some time before. Bathed in such a potent electron donor, aerobic organisms became experts at respiration, and this seems to have quickened the pace of biospheric metabolism. Ironically, the new aerobes created so much organic matter that this material could not be completely oxidized, and as it accumulated in aquatic environments the anaerobes of the Archean found themselves a new and trophically well-provisioned home.

Grypania, a spiral megascopic fossil from the time of the Oxygen Crisis, has been interpreted as the earliest eukaryote, but this assignment has been questioned because of the similarity of this elongate coiled form to gigantic bacteria or bundles of bacterial filaments resembling the modern sulfate oxidizing bacterium *Thioploca.* Reliable evidence for the presence of eukaryotes appears several hundreds of millions of years later. Acritarchs are hollow, microscopic, roughly spherical organic walled fossils removed from silicate rock with hydrofluoric acid. Acritarchs resembling those that might have been formed by modern sporopollenin cyst-forming protists first appear in rocks of China at 1.8 billion years ago. Small (but visible to the naked eye; 4–5 mm long) leaflike fossils 1.7 billion years in age, possibly remains of eukaryotes, occur in the Tuanshanzi Formation of the Changcheng Group in Jixian, China. These fossils, known as longfeng-shanids, are felt by their discoverers to be the earliest evidence of multicellular life. By 1.4 billion years ago, similar but often larger fossils of carbonaceous multicellular organisms (genera *Chuaria* and *Tawuia*) begin to become abundant in marine sedimentary rock localities throughout the world.

As the Proterozoic proceeds, more protoctists are added to the register of life, with many major groups making their first fossil appearance at or after 1 billion years ago. For example, a fossil somewhat resembling the modern red alga *Bangia* appears in Proterozoic strata of Arctic Canada. What relationship the emergence of these new types of eukaryotes has to the consolidation of the long lived supercontinent Rodinia is unclear. But with the breakup of Rodinia beginning in the late Proterozoic and ending at the Cambrian boundary (with the creation of Gondwana), tectonically driven geochemical changes of great magnitude occurred throughout the marine biosphere.

The world's worst ice age accompanied the breakup of Rodinia, approximately 750 million—550 million years ago (an interval that includes both the Sinian Period and the subsequent Lipalian Period). Some geologists see this glaciation as so severe that most of Mirovia (the superocean corresponding to Rodinia) was covered with sea ice. This great ice age began to abate some time before 600 million years ago. Perhaps not merely coincidentally, at 600 million years the first animals and the Ediacarans joined the biosphere. The appearance of the Ediacarans (a discussion of these forms will follow) marks the beginning of the Lipalian Period.

The earliest animal fossils are simple tracks and trails (trace fossils or ichnofossils) occurring in shallow marine sediments deposited after the great glaciation. At first the diversity of these trace fossils was low, but as the end of the Proterozoic is approached the diversity of traces increased dramatically. This is thought to reflect an increase in the behavioral sophistication of the animals responsible for the burrowing. Trace fossils have been described as "fossil behavior" (Lockley 2000), and although any given trace fossil cannot necessarily be matched with the phylum of animal responsible for the trace, these sediment markings do give a quite reliable impression of the behavioral sophistication of the organism that formed them. By the end of the Lipalian Period, combinations of burrowing patterns (the combination of a sinusoidal track and a spiral track, for instance) led to preserved burrowing patterns that are felt to have represented more sophisticated and efficient strategies for mining digestible organic matter from loose sediment (figure 3).

Associated with the earliest confidently identified animal trace fossils are the Ediacarans. These enigmatic creatures, ranging in size from a centimeter to even a meter, have been a focus of considerable paleontological controversy. Long held to represent the earliest animal body fossils, the Ediacarans are now thought by many to represent an unusual extinct phylum of life, a sixth kingdom to be added to the existing five (plants, animals, fungi, protoctists, bacteria). A strong case can be made for erection of this sixth kingdom (called Vendobionta), although a protoctistan assignment for the Ediacarans is defensible as well.

I have recently determined that the organization of the Ediacaran body plan must be understood as an example of metacellularity rather than of conventional multicellularity as seen in plants and animals. The larger Ediacarans developed by incremental addition of modular compartments called cell families. Iteration of an original number of one to six (sometimes more) cell families led to mature Ediacarans with body symmetries not often encountered in the animal kingdom. Perhaps the best-known example is *Pteridinium simplex*, an Ediacaran from Namibia with a ribbon-like body and a trifold cross-section. An undescribed Ediacaran (informally known as Vendofusa) from Newfoundland shows bipolar iteration of the cell families. This specimen falsifies the animal hypothesis of Ediacaran affinities, for it resembles a two-tailed worm more appropriate to the tale of Dr. Doolittle than to the world of real paleobiology.

The metacellularity model for the Ediacarans can also explain the bizarre symmetry changes that can occur in Ediacaran lineages. For example, the heraldically trifold Ediacaran *Tribrachidium* enlarged by growth of a three-cell-family cluster without iteration. It has a nearly exact counterpart in a two-cell-family Ediacaran, also without iteration.

Figure 3
The type section for the Lipalian Period, in the Cerro Rajón section of northwestern Sonora, Mexico. The boundary between the Lipalian and the Cambrian Periods occurs near the top of the ridge in the center of the photograph. The cliff forming unit on the horizon is the Cambrian Proveedora Quartzite.

This Australian species (*Gehlingia dibrachida*) is identical in details of its form to *Tribrachidium*; however, as a consequence of the fact that *Gehlingia* began life as a pair rather than a triplet of cells, the mature form of gehlingiids is bilaterally symmetric and frond-shaped. The radical difference in mature form in these two genera is a direct result of differences in the original number of adhering cell families.

The metacellularity model was recently confirmed by the discovery of a previously unknown bipolar Ediacaran with four-cell families in the Carolina Slate Belt. Discovered by Ruffin Tucker of Concord, North Carolina, an amateur fossil collector, the new form represents a pattern predicted by the metacellularity model. The predictive power of this model indicates that it is indeed the correct approach to interpreting Ediacaran growth.

Reproduction in Ediacarans may have been accomplished in a simple manner by splitting off of the requisite number of adhering cell families from the site of generation of new cell families. One could easily imagine the evolution of an iterated form from a noniterated form by failure of release of cell-family propagules. Instead of being released, these reluctant propagules (which refuse to leave the nest) are fused to their parental cell families.

With this advance in our understanding of Ediacaran body form, we can now address the ecological conditions of the Proterozoic. Ediacarans have bodies that would have been particularly well suited to passive strategies of acquiring trophic resources. Strategies such as photosymbiosis, chemosymbiosis, and direct absorption of nutrients from sea water accord with the high-surface-area somatic geometries of the Ediacarans. I have thus characterized the late Proterozoic marine biosphere as a "Garden of Ediacara," a time when collection of dispersed (rather than concentrated) food sources was emphasized in the members of the marine biota. Heterotrophy, although certainly present during the Sinian and Lipalian periods, took place mainly at microbial scales.

As the end of the Lipalian was reached, the greatest ecological transformation in the history of the marine biosphere occurred as the Ediacarans were displaced in shallow marine environments by diverse skeletalized animal communities at the Lipalian-Cambrian boundary, approximately 541 million years ago. These new communities included large (up to several meters in length) and apparently very aggressive predators such as *Anomalocaris*. Ediacarans rapidly vanished from the sea floor, probably because their light-seeking and high-surface-area-requiring life styles precluded the development of robust armor as a defense against large animal predators.

A few Ediacarans survived the boundary event to live on into the Cambrian. *Climactichnites*, a large enigmatic Cambrian fossil from sandstones deposited in the high intertidal zone, has long been interpreted as a trace fossil of a grazing mollusk-like animal of relatively large size. Curiously, however, some of the *Climactichnites* "traces" appear to have been eaten or otherwise intentionally disturbed by animals, and specimens are known in which the end of the "track" tapers distally to form an imbricate set of elongate units very reminiscent of the cell-family-generating tip of the frondose Australian Ediacaran *Charniodiscus*. Thus, it may now be time to revive the old idea that *Climactichnites* is a body fossil rather than a trace fossil. If so, it may represent the last (and largest) Ediacaran, driven into a marginal environment (the high intertidal zone), which would then represent the final refugium for Ediacarans living a Garden of Ediacara life style.

In the foregoing I have been indulging in a style of historical reconstruction that is very much a part of the Western scientific tradition. In fact, its origin can be traced back to the disciplined research program begun by the great paleontologist Georges Cuvier. With regard to the past behaviors and autecologies of extinct organisms, Cuvier wrote the following in 1798: "One could even, with a little more boldness, guess [*deviner*] some of its habits; for the habits of any kind of animal depend on its organization, and if one knows the former one can deduce [*conclure*] the latter. After all, these conjectures would hardly be any more hazardous than those that geologists are going to find themselves obliged to make. . . . " This type of conjectural approach, involving the generation and testing of multiple hypotheses, is antithetical to the Eastern scientific approach advocated and championed by Vernadsky. Vernadsky wrote in 1926 that such approaches "hinder scientific research by limiting its final results; by introducing conjectural constructs based on guesswork (*ugadyvat'/ugadat'*), they obscure scientific understanding." Although Vernadsky's cautionary comments are certainly worthy of note, it seems clear to me that if we are to learn more about life from the interval between its origin to the end of the Lipalian period we must proceed with a significant component of the Western, Cuvieran approach to the scientific analysis of these events of the remote past.

Readings

Butterfield, N. J., A. H. Knoll, and K. Swett. 1990. A bangiophyte red alga from the Proterozoic of Arctic Canada. *Science* 250: 104–107.

Lockley, M. 2000. *The Eternal Trail: A Tracker Looks at Evolution*. Perseus.

McMenamin, M. A. S. 1998. *The Garden of Ediacara: Discovering the First Complex Life*. Columbia University Press.

McMenamin, M. A. S., and D. L. S. McMenamin. 1990. *The Emergence of Animals: The Cambrian Breakthrough*. Columbia University Press.

Rudwick, M. J. S. 1997. *Georges Cuvier, Fossil Bones, and Geological Catastrophes*. University of Chicago Press.

Scamardella, J.M. 1999. Not plants or animals: A brief history of the origin of kingdoms Protozoa, Protista, and Protoctista. *International Microbiology* 2, no. 4: 999.

Shixing, Z., and C. Huineng. 1995. Megascopic multicellular organisms from the 1700-million-year-old Tuanshanzi Formation in the Jixian area, north China. *Science* 270: 620–622.

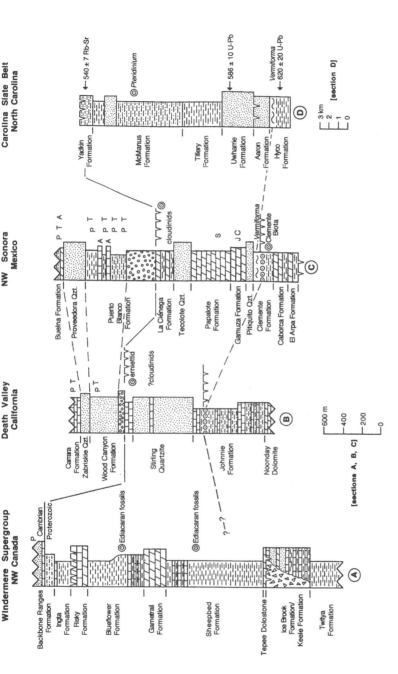

Figure 4

Correlations among four key Lipalian stratigraphic sections of the western hemisphere: (A) Windermere Supergroup, northwestern Canada; (B) Death Valley, California; (C) northwestern Sonora, Mexico; (D) Carolina Slate Belt, North Carolina. Radiometric dates are in millions of years. A: archeaocyaths. C: conical stromatolite *Conophyton*. J: branching stromatolite *Jacutophyton*. P: phosphatic shelly fossils. S: domal stromatolite. T: trilobites. Adapted from M. McMenamin, Ediacaran biota from Sonora, Mexico, *Proceedings of the National Academy of Sciences* 93 (1996): 4990–4993. Added in proof. For details, see original paper.

10   Continental Drift and Plate Tectonics

Raymond Siever

In this chapter, Raymond Siever traces the development of the discoveries and ideas that led to the synthetic theory of the workings of the Earth's surface, the grand unifying concept of geology: plate tectonics. Echo soundings and magnetometer tracings of the deep ocean, the distribution of volcanoes and earthquakes, paleontological studies of fossil reptiles, and other disparate sources of information reveal a new view of our dynamic Earth. Siever is a professor of geology at Harvard University.

The old ideas that oceans and continents were permanently in their present positions were rooted in our emotional need for something firm in our lives: the Earth beneath us, always still and always stable. People living in earthquake zones always knew that the Earth shook, yet it didn't seem to move around between earthquakes. The concept of stabilism, taken for granted at least since the early 1800s, was assumed in geology as a consequence of a very important doctrine: uniformitarianism. Uniformitarianism, enunciated early in the nineteenth century, states that processes at the Earth's surface operated in the past as they operate today. Rivers excavate valleys as rivers always excavated valleys in the past. Although not specifically written, it seemed obvious that if past geological processes always worked the way they do now the continents must also have always been in the same positions.

All this has changed now. The comprehensive theory of plate tectonics has kept geologists busy since its synthesis in the late 1960s. It is odd for me, having been involved in plate tectonics from the beginning as an onlooker, to consider plate tectonics now as an old theory, as old as some ideas about Earth that held sway before. Earlier "theories" were not really scientific theories. They were fanciful notions, like "the cooling Earth contracted like an orange, and mountains were created when the orange peel was crenulated and ridged." Until about 1965 no one had good ideas of why or how the Earth's surface behaves. Mountains, volcanoes, and the distribution of earthquakes were known—we had explanations of how these features acted, but no concepts to tie them together. The discovery of coal beds in the far north at Spitzbergen and in Antarctica disturbed the standard geological world view. Nothing like subtropical forests can grow in these climates now, yet such coal beds must have come from swamp vegetation that grew under hot and humid conditions.

Evidence for Drifting of the Continents

"Continental drift" was one of the earliest ideas people had after looking at decent maps. This idea may have already occurred to Francis Bacon, who in 1620 commented on the fit of Africa and South America. In 1858, Antonio Schneider, a mapmaker, constructed a map of the Atlantic Ocean showing how South America and Africa may have drifted apart. But the true father of continental drift was Alfred Wegener, a German who studied world climatology. Wegener helped discover the tropical *Glossopteris* flora of the southern hemisphere. He questioned why glaciations had occurred at times in the past on continents that are now tropical. The ideas of Wegener (and, later, of others who believed in continental drift) were not accepted by most geologists. Wegener was an outcast because he wasn't a "true geologist"; his ideas were also rejected because he simply had no decent mechanism for "continental drift."

Wegener presented the first serious challenge to stabilism when he noted not only that the contours of South America and South Africa fit but also that the fossils of the west coast of Africa matched those of the east coast of South America. There was, for example, fossil evidence that a small late-Paleozoic reptile known as *Mesosaurus* lived a little more than 250 million years ago in what is now South America and Africa. Wegener thought it exceedingly unlikely that *Mesosaurus*, a freshwater swimmer, would be limited to those two continental locations without having spread to many other environments. If it could swim across the Atlantic Ocean, why did one find *Mesosaurus* only at a few specific locations (figure 1)? Indeed *Mesosaurus* was not much of

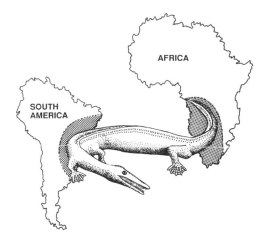

Figure 1
The distribution of fossils of the dinosaur *Mesosaurus* on the west coast of Africa and the east coast of South America. *(drawing by Sheila Manion-Artz)*

a swimmer; it probably could not have crossed the ocean. Wegener reasoned that *Mesosaurus* fossils were found in rocks of the same age at these disparate places because, in the late Paleozoic, these two continents were together. Wegener generated many other biological arguments based on his proposal that the continents, known from geophysical evidence to be approximately 25–30 kilometers thick and overlying a much denser substratum of the Earth, drifted in the oceanic crust like floating pieces of ice in the sea.

The defenders of stabilism responded almost immediately. Biologists and paleontologists invented ingenious alternative explanations for the distribution of fossils. The example of *Mesosaurus,* for example, was explained by the supposed existence of a sunken and eroded land bridge across the South Atlantic Ocean. Wegener argued that continental material is so different from oceanic crustal material in density and composition that a sunken land bridge would be detectable. Geophysicists, for their part, rejected any process of continental movement; they could not conceive of any dynamic mechanism by which continents 25–30 km thick could plow through oceanic crust. Without a mechanism, Wegener's concept of continental drift remained at the periphery of mainstream science.

A general theory of the formation of mountains was lacking as recently as 1960. Geologists believed in continental accretion—that continents grew steadily during geologic history but always stayed in the same place with respect to the ocean basins. The first scientist who proposed a reasonable mechanism was Arthur Holmes, a famous geologist from Scotland. In the 1920s Holmes proposed that the mantle, the interior of the Earth, convected heat, like hot coffee rising unevenly to the top in a just-poured cup. The mantle, beginning approximately 30 km down in the interior of the Earth, was known to be very hot relative to the cooler outer part of the Earth, the crust. Large convective forces transmitting heat from below to above would set up "convection cells" responsible for mountain production.

The Earth's convection, as any convective current, is the result of a temperature difference in the gravity field. The hotter the material is, the lower its density. The material that heats up in some portion of the hotter mantle will become less dense as a result of increased temperature and will tend to move upward, seeking and finding its own level of density. The cooler material tends to sink. This is the behavior of any convection cell.

Holmes held that where the crust and the mantle meet, downbows in the Earth's crust would be responsible for the first building of geosynclines believed to cause continental accretion. (Geosynclines were defined as linear belts of thick accumulations of sediment that were later deformed into mountain belts.) Sediments would accumulate at continental edges: as motion continued, crumpling, so said Holmes, would erect large mountain belts, introducing great bodies of molten rock appearing on the surface as volcanoes. Lava extrusion, deformations resulting in mountains, Holmes suggested, could be accounted for by convective cells. Holmes's good idea was not easily tested; it rested for 30 years until the period of ocean exploration that started just before World War II.

Mapping the Sea Floor

The major development that led to the measurement of ocean topography, echo sounding, started in the 1920s. Ships on the ocean's surface sent down pulses of sound, which were reflected back up to them. From the travel time of these sound waves and an accurate knowledge of the velocity of sound in water, the distance to the bottom was deduced. Every ship with an echo sounder detected subsea hills. A topography of an ocean floor, typified by the Atlantic Ocean, was revealed (figure 2). From sea level down to about 200 meters is a broad,

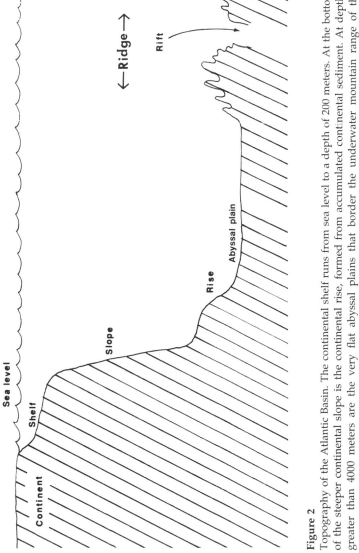

Figure 2
Topography of the Atlantic Basin. The continental shelf runs from sea level to a depth of 200 meters. At the bottom of the steeper continental slope is the continental rise, formed from accumulated continental sediment. At depths greater than 4000 meters are the very flat abyssal plains that border the underwater mountain range of the mid-Atlantic Ridge.

shallow apron bordering the continents (the *continental shelf*); it is followed by a steeper slope of about 5° (the *continental slope*). Below the slope is the *continental rise*, a great pile of sediment derived from continental erosion settling out into deeper ocean waters. Further out, at depths greater than about 4000 meters, are smoother expanses called the *abyssal plains*. These plains give way farther out to a hilly topography that culminates in a high ridge of mountains, the mid-Atlantic ridge, running down the middle of the Atlantic Ocean. In the Pacific Ocean the same general pattern was found, plus another kind of topography: some abyssal plains or hills are bounded by very deep trenches, some 10,000 meters below sea level.

The discoveries that led up to the theory of plate tectonics occurred around 1965—mostly as results of the work of Maurice Ewing and his associates at Columbia University's Lamont-Doherty Geological Observatory, who mapped the ocean floors extensively in collaboration with scientists from other oceanographic institutions. A relatively recent map (figure 3) reveals mountains, valleys, plains, and the peculiar topography of the sea floor surrounding the continents. The sea floor has mountain ranges similar to the Alps and the Himalayas but arranged somewhat differently. The work corroborated the suspicion of the Challenger Expedition of the late nineteenth century: the mid-Atlantic ridge is a series of high mountains running down the center of the Atlantic Ocean. The center of this high ridge of mountains is a deep valley: a symmetrical cleft or crack running along the middle. We now know that these mid-ocean ridges (MORs) are distributed around the globe. The mid-Atlantic ridge, for example, can be traced into the Indian Ocean. Ridges cross the other oceans too. The East Pacific rise is like the mid-Atlantic ridge but lower and broader.

The explanation of these ridges was at first a mystery. What did these cracks signify? The ridges were receiving very little sediment: they were floored with only a thin layer. Beneath the thin layers of sediment was a rough surface, opaque to any further transmission of sound. Some rock down there seemed to be much denser; it was opaque and would not transmit anything. Signals coming back from this "layer B" were later identified as a worldwide substratum of basalt. Layer B could be dredged, and pieces were broken off from the top; it outcropped in part of the abyssal hills and then in the mid-ocean ridges.

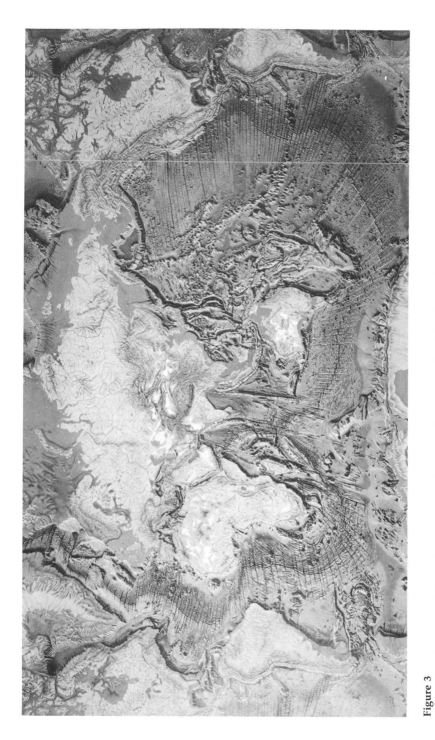

Figure 3
A map showing some of the underwater mountain ranges that traverse the globe. (*From* The Floor of the Oceans, *based on bathymetric studies by B. C. Heezen and M. Tharp. Copyright 1977 Marie Tharp. Reprinted with permission.*)

Putting together the detailed discovery of mid-ocean ridges with the discovery of basalts in the 1960s, Harry Hess of Princeton University enunciated the idea of sea-floor spreading. Hess, a mineralogist who had been a submarine commander during World War II, became involved in the history of the oceans. He had worked on the history of the Caribbean, still an inexplicable geological problem. Hess came up with the idea that the mid-ocean ridges had been created as the deep interior of the Earth welled up, producing molten basalt. Basalt, a dark igneous rock, is the product of lava flows. One of the most abundant rock types on Earth, it constitutes much of the lower part of the crust everywhere beneath the continents. It erupts at the surface in a great variety of volcanoes, typified by Kilauea and Mauna Loa on the Hawaiian Islands. Basalt contains many small pieces of a glassy material and very fine fragmented crystalline materials, which suggest that it is formed by the rapid quenching of a very hot igneous rock melt or magma. The temperatures of volcanoes underlain by basaltic lava have been measured as 1000–1100°C. When basaltic material is spewed high into the air as volcanic eruptions of the explosive type, or when it pours out onto the surface as lava streams, as observed on Kilauea, it quickly cools, forming glass; thus, it offers little opportunity for systematic and orderly growth of crystals. Basalt outflows have been known all over the continents from regions of vast upwellings, typified by the Snake River and the Columbia Plateau of the state of Washington, in which thousands and thousands of square miles are covered by dark basalt lavas that flowed out in the last 50–60 million years. The central fissure eruptions of Iceland are basaltic. This rock type, which is present in so many different areas of the Earth and is known to be one of the lower layers of the crust, assumes special significance when it wells up in the mid-Atlantic and other mid-ocean rises to form the new parts of the crust. Basalt is the material that comes out of the Earth's interior to make the new parts of the plates.

As basalt wells up, the ocean floor splits, rifts, and then continues to spread apart. Hess's problem was urgent: "Where did the floor of the ocean go?" Many of us accepted Hess's mechanism of sea-floor spreading at the rift almost as soon as he presented it. But what happened to the old rock? We could not believe that the Earth was expanding. The expanding-Earth hypothesis, periodically proposed, was always demolished—with good reason. The search was on.

Hess combined Arthur Holmes's ideas of convection currents with the new ocean data: a convective upwelling of basalt was formed by the rise of deep mantle material. The central valley was a rift; material moved apart and out at this point. Hess envisaged a giant "conveyer belt" bringing up material from deep inside the mantle that gradually spread out farther from mid-ocean centers. I went out in 1965 to the mid-Atlantic ridge, at about 20° north latitude, to study sediment basins on and alongside the ridge. Very thin layers of sediment were near the ridge; the sediment became thicker the farther out one went. We also discovered manganese oxide being deposited at a slow rate in the form of crusts that became thicker as one moved away from the ridge. These discoveries were consistent with the idea that the hills farther away from the rift were older. As the ocean floor was made by basalt coming up at the rift, the conveyer belt operated and materials moved away.

Magnetic Anomalies: Clues to Sea-Floor Spreading

Then came the crucial explanation of sea-floor spreading, which was based on the discovery of magnetic anomalies. Fred Vine and Drummond Matthews of Princeton University set this picture in order. Magnetometer readings from several different traverses taken at different places in the ocean could be lined up. In its simplest form, a magnetometer is a magnetic compass needle, suspended by a fiber in the Earth's magnetic field. A bar magnet of known strength is brought within a known distance of the compass needle. Knowing the strength of this magnet, we calculate the attractive force. This attractive force will be countered by the force of the Earth's magnetic field attracting the compass to the north. The needle will be deflected toward the magnet, but not as far as it would be if the Earth's magnetic field were not there. The fundamental principle of how a magnetometer works is that it simply compares the Earth's magnetic field with a known magnet of some kind to deduce the strength, at that site, of the Earth's magnetic field. The most sensitive magnetometers available now are ones that depend on the atomic magnetic properties of the proton.

Vine and Matthews discovered that the curves of the strength of the magnetic field along different traverses could be lined up. These anomalies formed striped patterns, which covered the ocean floor.

Although they didn't know what caused these anomalies, Vine and Matthews established that the patterns were produced by magnetic reversals. Every so often in the Earth's history the magnetic field suddenly flipped: where there had been a north-seeking pole, the magnetic field became south, and vice versa. Vine and Matthews discovered that layers of rocks of different polarity alternated.

These patterns can be explained by the dynamics of the sea floor (figure 4). Imagine magma coming up from the hot part of the mantle. The dark basaltic lava flows from the partial melting zone out onto the sea floor. As it cools into rock, the temperature of the material descends from roughly 1000°C to about 500°C. Tiny amounts of iron in the rock act as little magnets. As the lava crystallizes below 500°C (which is called the Curie point, in honor of Marie Curie), the magnetism is locked into the rock in such a way as to record the direction of the Earth's magnetic poles at the time the rock congealed. As long as the polarity of the Earth's magnetic field remains the same, the magma retains the same signature. Operating much as a tape recorder does, the ocean floor preserves a record of the Earth's magnetic field, recording the magnetic orientation in the basalts or lavas as they flow out on the surface. The lavas are carried out by seafloor spreading. The lavas that are magnetized thus carry a record of the direction of the magnetic field. In addition, they are a taped record of the time when they were deposited, and thus of how fast the sea floor was spreading. Before Vine and Matthews, most magnetic stratigraphy measurements were made on land.

The time scale of magnetic reversal (figure 5) shows that the present polarity has existed less than a million years. Then there was a short period at the beginning of the Matuyama reverse period when north was south and south was north. Flip-flops occur on an irregular time scale of more than 3 million years, beginning well over 100 million years ago. We have only recently begun to find out why reversals occur.

We know from steady and consistent measurements made at magnetic observatories that the Earth's magnetic field has been decreasing in the last 140 years. We are probably on the way to another magnetic reversal, which will occur as the present magnetic field decays down to zero and then builds back up again with a reverse polarity. Reversals take place every 10,000–100,000 years or so; thus, we can

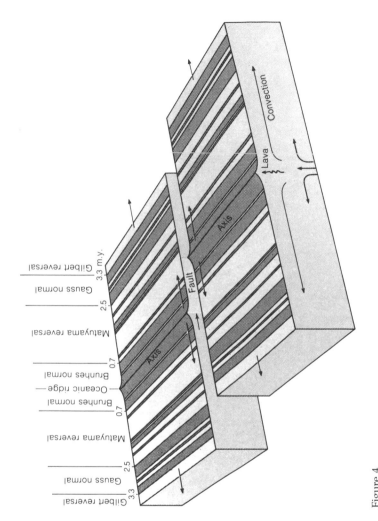

Figure 4
Symmetrical patterns of magnetism on either side of a mid-ocean range. (*From F. Press and R. Siever, Earth, fourth edition (Freeman, 1986). Copyright 1986 W. H. Freeman and Company. Reprinted with permission.*)

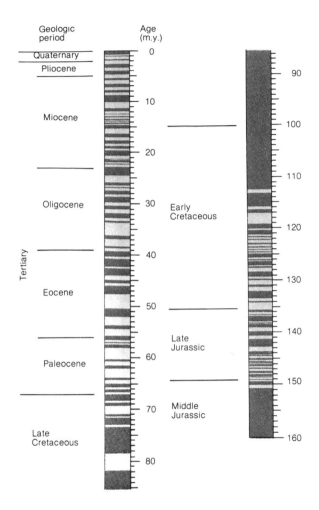

Figure 5
A time scale, stretching back from the present to the Jurassic, based on magnetic
reversals. The darker bands represent periods of so-called normal polarity, when the
North and South magnetic poles corresponded to their present positions; the lighter
bands represent periods of reversed polarity. *(From F. Press and R. Siever,* Earth, *fourth
edition (Freeman, 1986). Copyright 1986 W. H. Freeman and Company. Reprinted with permission.)*

confidently expect one to happen again sooner or later, particularly since we have been in a normal era for many thousands of years. During the reversal, as the magnetic field declines to zero, magnetic compasses will be confused, for there will be no north or south pole to attract the compass needle. Since the magnetic field acts to deflect some of the radiation impinging on the Earth from outer space, we may temporarily get more radiation. Yet the history of life suggests that during the many reversals no conspicuous effects on the nature of life occurred.

By early 1965 another clue to the tectonic puzzle had been found: paleomagnetism. The poles "wandered" (figure 6). Nearly all rocks retain a small amount of the magnetism that is induced at the time of their formation. If we look carefully at the magnetic signature of a lava that is approximately 10 million years old, we discover that the north magnetic pole was then in a different place than it is now. If the rock contains a record of the ancient magnetic pole positions, then we can take a great many rocks formed at a great many different times and note how the poles' positions seem to have changed over time. Now fathom this: If all the continents were in exactly the same positions relative to one another then as now, all polar wandering paths should be identical, there being at one time only a single magnetic field with one north and one south pole of the Earth. No theory suggests any magnetic field that involves more than just the north and south poles. The fact that the data yield different "polar wandering paths" means that the continents have moved with respect to one another. In the 1960s, in the midst of all the ferment about an extremely active sea floor, these precise "polar-wandering curves" really upset the apple-cart. Geophysicists started considering ideas of moving continents, even though geologists were still a little bit reluctant.

Disappearance of Crust at Subduction Zones

In 1965, Hess's original question still needed answering: If new basalt crust is appearing at the mid-ocean ridges, and the earth is not expand-ing, where does the crust disappear? The oceanwide, worldwide earth-quake-sensing network, set up mainly to monitor seismic results of underground atomic explosions, made it possible for the first time to understand the distribution of earthquakes. When the map reproduced

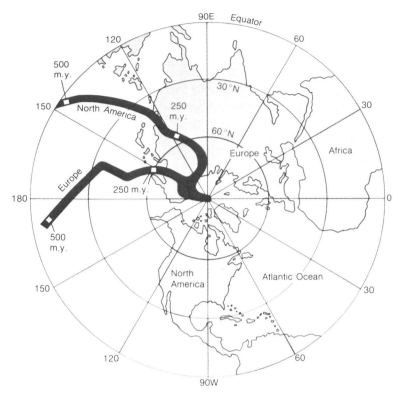

Figure 6
Polar wandering curves of North America and Europe. *(From F. Press and R. Siever, Earth, fourth edition (Freeman, 1986). Copyright 1986 W. H. Freeman and Company. Reprinted with permission.)*

here as figure 7 first appeared, it was considered remarkable that all the rift-ridge and trench zones under the oceans were outlined. An enormous volcanic ring of fire consisting of all the volcanoes around the Pacific Basin was revealed. Earthquakes outline the mid-Atlantic ridge between the Americas and Europe and between South America and Africa; they also delineate the area between Africa and Antarctica. Why are earthquakes patterned like this?

The tearing that gives rise to earthquake waves takes place at varying depths, some as deep as 700 km. Shallow-focus earthquakes characterize the mid-Atlantic ridge and the East Pacific rise running from North America southward to the coast of South America. The deep-focus earthquakes are typically in the trenches on the western side of

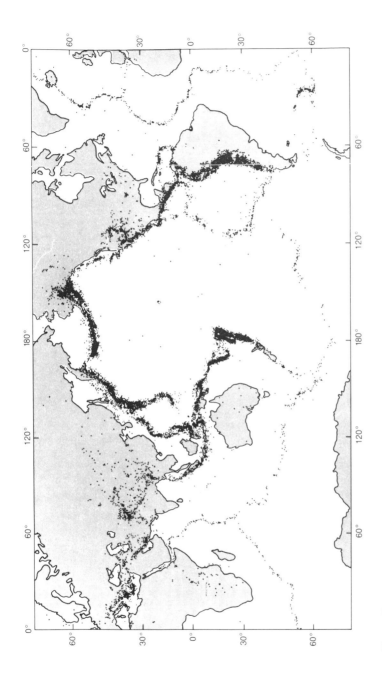

Figure 7
Distribution of earthquakes recorded during the period 1961–1967. *(From F. Press and R. Siever, Earth, fourth edition (Freeman, 1986). Copyright 1986 W. H. Freeman and Company. Reprinted with permission.)*

the Pacific Ocean. Deep-sea trenches are great deep folds in the oceanic crust. Lynn Sykes established that in the trenches the earthquakes are distributed in a descending zone, going deeper and deeper as one moves from the trench toward volcanic islands associated with the trench. The trenches are areas of abnormally low heat flow, whereas over the ridges (such as the mid-Atlantic ridge and the East Pacific rise) heat flow is much higher. The high heat flow results from upwardly moving convection currents transmitting heat from deeper in the mantle to the surface. This implies that in the trenches there is a downward movement of cool material, since this is an area of low heat flow relative to the heat flow of the ridges.

The fact that trenches are areas of downward movement of cool crust and the descending distribution of earthquakes indicate that oceanic lithosphere is sliding down the trench, moving deeper into the mantle. The places where this happens are *subduction zones.* That subduction zones are where the plates go was realized by Dan Mackenzie and too many other people to name in late 1966 and early 1967. The final picture of plate tectonics fell into place: the ocean ridges are where basalt comes up, creating the lithospheric plates, which move out as rigid bodies until a subduction zone is met. At the subduction zone, the crustal plates start to dive down, completing the cycle.

Movement of Lithospheric Plates: The Essence of Plate Tectonics

A final clue that made everything fit into place was the concept of lithospheric plates, varying in thickness from 70 to 100 km. Below the plates lies the asthenosphere, defined as the partially plastic upper part of the mantle. The asthenosphere can flow and deform by plastic flow, whereas lithospheric plates behave as rigid bodies. This view of the Earth came primarily from a detailed study of the transmission of seismic waves. The Earth's surface consists of rigid plates riding on a relatively plastic and movable asthenosphere. By matching the North American, South American, and African continents and the western part of the Eurasian continent plus Greenland, one can obtain the best possible fit using the continental-shelf lines—rather than the shoreline—where properly speaking, the continents end. Wegener did this—but now we understand the mechanics. The great cry of the geophysicists in the 1920s was that continental drift couldn't work because a continental crust couldn't plow through the oceanic crust in even a shallow ocean. Now we realize that constituents of the lithospheric plates, about 100 km thick, do not plow through oceanic crust; they simply ride over a plastic asthenosphere. The continents are simply the uppermost portions of the moving lithospheric plates, and the oceanic crust is produced at the trailing end of the plate and consumed by subduction at the leading end. The paleomagnetism and the magnetic time scale tell us when and how rapidly the plates move. The rate of sea-floor spreading may be as small as 2–4 centimeters per year. For example, the rift in the center of Iceland is moving apart at an annual rate of about 2 cm.

Between 1965 and the spring of 1967, new ideas floated around the geological, geochemical, geophysical, and oceanographic communities. The new geological ideas led to an explosion of papers in 1967 and 1968 by half a dozen geophysicists, all of them very young and active in oceanography and geophysics: Xavier LePichon, a French geophysicist working at Lamont-Doherty Geological Observatory; Lynn Sykes, a seismologist at Lamont-Doherty; Dan MacKenzie, a geophysicist from Cambridge, England; Jason Morgan from Princeton; and many others. Each was working furiously on a piece of these developments. They visited one another's laboratories, jumped on and off ship, and flew to other ships. The whole picture then emerged. Tested and confirmed in so many different areas, "plate tectonics" is now completely accepted as the major driving force of the dynamics of the Earth.

To understand plate tectonics we must consider the structure of the outer zones of the Earth. The thin outer crust of the Earth is cool relative to the hot interior beneath both the oceanic crust and the continental crust. Continental crust is somewhat thicker than oceanic crust; each is part of the lithosphere. The lithosphere comprises the outer layer of rock, which is more or less rigid, solid, and brittle. Below the lithosphere is the darker zone of partial melting, the top of the asthenosphere. Pressures and temperatures, particularly in the zone of partial melting, are so high in the asthenosphere that its material deforms as a plastic solid. As any other fluid, the asthenosphere's material is characterized by a viscosity, but the asthenosphere is 22 or 23 orders of magnitude more viscous than the fluids with which we are accustomed. The materials of the asthenosphere move very slowly. The asthenosphere is heated by sources inside the Earth—primarily by the incorporation of radioactive uranium and thorium minerals, and a little-known but very important radioactive isotope, potassium 40. These materials all contribute to the heat of the Earth, which is slowly being transmitted to and through the surface. The interior of the Earth is not all molten, although the core is. For years geophysicists worried about the heat flow because so much heat from the Earth could not be transmitted to the surface by heat conduction. (Heat conduction is a familiar phenomenon; if you sit on a radiator, you get hot because of direct conduction of heat from the heat source to you.) Thermally, the Earth is a lot like a brick: it takes a long time to heat up, and it takes a similarly long time to cool off; thermal conductivity is extremely slow. Almost 100 years ago it was shown that the Earth produced internal heat by radioactive decay at such a rate that it could not possibly lose it all only by heat conduction. Another major way of losing heat is by convection, the way a radiator heats a room: The hot air rises because it is less dense; in some other part of the room, the dense, colder air sinks to the bottom. The easiest way to see convection currents is in a cup of coffee with the proper angle of illumination of light coming in at the borders. These convection "cells" constantly form and break up as the coffee cools from the top layer and then heats up from the hot interior.

The mantle turns over constantly, losing heat in the same way that a room loses its heat in winter even though the radiator continually pumps heat into the room. The cool upper surface of the mantle is at the boundary where the oceanic crust meets the lithospheric crust. As the mantle convects, it brings material to the top, where it cools off more directly than it would by heat conduction throughout the great thickness of the mantle. Hot material from the zone of partial melting in the mantle is brought from the interior to the surface and extruded on the surface at mid-ocean ridges, where it quickly quenches under seawater or cools if it comes up on land (as in Iceland, where the mid-Atlantic ridge passes through the large island).

Plate tectonics even explains how Earth keeps the same radius even though new ocean floor is constantly being formed. The idea that Earth has a constant mass is old; it has not been changed by plate tectonic theory—rather, it supports the concept that plate tectonics is not the result of an expanding Earth. The analysis of Earth's constant radius derives from its constant mass. Not much of Earth's material is lost to space. Hydrogen gas does have sufficient escape velocity to leave the atmosphere. Small amounts of helium may leave by gaseous escape from the outer atmosphere, but Earth's gravitational pull is so strong that few other materials leave. The planet has a constant mass, for practical purposes, and the value of gravity always has been about what it is today. The gravitational compression pulling the planet together accounts for the high density and temperature of its interior. Earth long ago reached equilibrium with respect to gravity, which holds it together. The sun, since it has nuclear fusion in its interior, could conceivably expand; however, no such nuclear mechanism exists for the much cooler Earth. Therefore, if material is produced on the sea floor, we must find a place to which it returns. Plate tectonic theory solved the problem of where the lithosphere, coming up and created at mid-ocean ridges, disappears. The answer is at the subduction zones.

The basalt, cooled from the ascending magma, comes up, and the sea floor, the lithosphere, moves out from each side of the rift (figures 8 and 9). At the subduction zone, crust dives down into the mantle where it is resorbed. The cool lithospheric plates go down into the trench. The material cools down as it travels along. The cool, water-logged sediments enter a deeper zone of the mantle (figure 10; left and right in figure 9). They enter a zone where materials start melting from the top layers, work their way up, and ultimately form volcanoes.

Several kinds of subduction zones exist. One is the Marianas type, which formed the island arcs of the Pacific; another is the Andean type, which created the Andes Mountains of South America. As a subducting plate dives below continental crust, the melting rock comes up underneath, creating huge volcanic mountain chains like the Andes. Where plates are subducted underneath thin oceanic crust, island arcs such as the Marianas, the Tongas, and the Japanese islands form.

Transform faults are the accommodations the Earth makes when plates slide past each other (figure 11). In mid-ocean ridges a transform fault appears as a crack and a lateral movement. Although they were understood before the advent of the theory of plate tectonics, transform faults were quickly included in the overarching concept. When mid-ocean ridges curve around and change direction, they tend to tear along such transform faults. For example, the San Andreas fault in California, which was responsible for the 1989 San Francisco earthquake, is a large transform fault. The mid-East Pacific rise, running south from Baja California under the continent, shifts so that the Pacific plate moves laterally northwest relative to the North American plate;

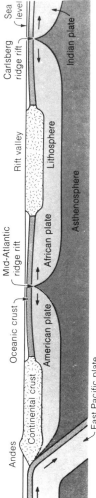

Figure 8

The American plate and the African plate are moving apart at the mid-Atlantic ridge. The East Pacific plate is being subducted under the western edge of the South American plate, creating the Andes. *(From F. Press and R. Siever, Earth, fourth edition (Freeman, 1986). Copyright 1986 W. H. Freeman and Company. Reprinted with permission.)*

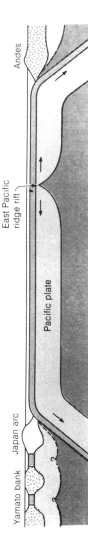

Figure 9

The East Pacific Plate, moving away from the East Pacific ridge, is subducted under thin oceanic crust, creating the island arc of Japan. *(From F. Press and R. Siever, Earth, fourth edition (Freeman, 1986). Copyright 1986 W. H. Freeman and Company. Reprinted with permission.)*

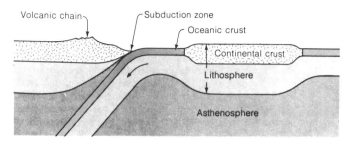

Figure 10
Subduction zones occur when oceanic and continental plates converge. *(From F. Press and R. Siever,* Earth, *fourth edition (Freeman, 1986). Copyright 1986 W. H. Freeman and Company. Reprinted with permission.)*

California is moving about a centimeter or two per year. As the Pacific plate moves to the northwest it slides past the North American plate, yielding many earthquakes that are an order of magnitude greater than the earthquake of 1989. These movements will continue as the plates move. "Baja" California will move up opposite to the interior of "Alta" California. Eventually Los Angeles will be pushed opposite Berkeley, with results we cannot predict.

How permanent is the current configuration of continents and oceans? We can construct a map of how the plates of the world are divided and in what directions they are moving (figure 12). The African continent seems firmly anchored, but the Atlantic Ocean grows wider by about 2 cm per year. The East Pacific rise moves apart much more rapidly.

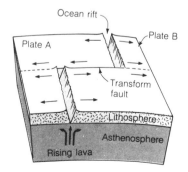

Figure 11
Transform faults occur when plates that are separating slip past one another.

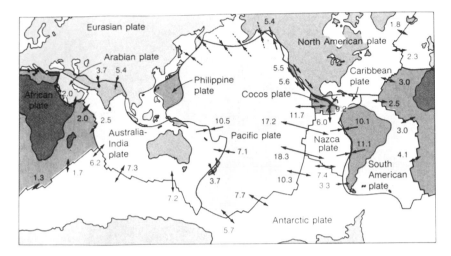

Figure 12
Modern plate boundaries. Opposed arrows indicate converging plates. Diverging arrows represent spreading at ocean ridges. Relative velocities of plate movements are given in centimeters per year.

Volcanic lava continues to come up at "hot spots," such as the Hawaiian Islands and Iceland. Hot spots may be controlled by the same forces that move plates, but we are still not sure. Hot spots are plumes of magma coming from perhaps as deep as the boundary between the mantle and the molten core, near the center of the Earth—much deeper in the mantle than the partial melting zone. Hot spots are fixed locations: as a plate moves across a hot spot, it leaves a trail of volcanoes. The Hawaiian Islands, at the end of a trail of a series of extinct volcanoes, were created as the Pacific plate moved northwestward across a hot spot.

Jason Morgan of Princeton University realized that hot spots had to somehow connect to plate tectonics. For the first time in the history of geology, it was felt that everything about the Earth's surface should be included in plate tectonic theory: the sea floor, mountains, volcanoes, volcanic island arcs, the Hawaiian Islands, and so on. And we now can explain continental drift: Continents happen to ride on the plates. The plates, not the continents, are drifting. The continents, as elevated portions of the plates, are passengers.

Glaciers, too, fit in beautifully with plate tectonics. Active glaciers now are limited to extremely high mountains or to polar regions.

But certain rocks containing coal beds must have moved; formed in tropical, humid climates, they are now in glaciated regions. Past glaciations inferred from glacial features now in the tropics must have originated when these continents were closer to the poles.

Reconstructing Past Movements

Plate tectonics has inspired research into how mountains were created, into paleo-plate tectonics, and into a topic important for evolutionary biology: extinctions and the disappearance of hordes of phyla at the end of the Paleozoic era. These extinctions can be attributed to the restriction of coastal and continental shelf areas when the continents were together.

The example of paleo-plate tectonics and Earth history helps spell out how we now use plate tectonics to reconstruct mountain-building events. During the development of a passive plate margin, upwarping occurs where molten rock comes up from below, bows up continental crust, and splits it apart into a rift valley. The rift valley widens and deepens; then the sea may invade. The Red Sea and the Gulf of Aden are examples of such valleys. The Dead Sea is a candidate for future marine invasion. The rift widens, seawater penetrates, and the continental margin cools. As it cools down, the plate contracts, and the margin gradually receives eroded sediment from the continents and thus subsides. The full development of a passive ocean margin is typical of the Atlantic Ocean, where extensive continental shelves delimit former plate boundaries. At active margins, Andean-type subduction zones, oceanic crust dives beneath the continental crust (figure 13). When a plate carrying a continent dives below another plate carrying another continent, the continents collide and are not subducted. The Himalayan Mountains were created in this way when the Indian subcontinent collided with the Eurasian plate. The processes occurring at active plate boundaries account for the development of most of the mountain belts of the world.

How do we reconstruct continental drift through geologic time (figure 14)? Turning the clock back, we see the Atlantic Ocean narrowing. Moving back 120 million years, we see South America and Africa joined. At 180 million years ago, North America also joins. By

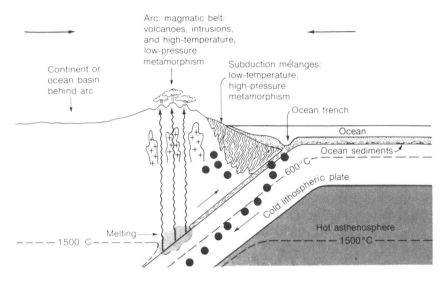

Figure 13
Andean-type subduction zone, an example of active plate margins. Collision of an oceanic plate with continental crust results in formation of trenches, accompanied by metamorphism, volcanism and earthquakes (represented by black dots). *(From F. Press and R. Siever,* Earth, *fourth edition (Freeman, 1986). Copyright 1986 W. H. Freeman and Company. Reprinted with permission.)*

240 million years ago, the configuration of the giant supercontinent, Pangaea, is formed.

The names Pangaea, Laurasia, and Gondwanaland were invented by the advocates of continental drift. In 1915 Alfred Wegener, the promulgator of continental drift, named Pangaea. The modern continents put back together give us the enormous ancient central continent combining South America, Africa, North America, Asia, Europe, Australia, and the Antarctic. Pangaea was Wegener's creation. He also used the term Gondwanaland, invented around 1885 by Edward Seuss, a Swiss geologist. Seuss had noticed, as had others, that fossils of certain distinctive land plants, the *Glossopteris* flora, occurred in the Gondwana Formation of India. This flora was also characteristic of fossils of the same age in South America and Africa. Seuss assumed that peninsular India was once somehow connected to the other southern continents. No one at that time knew about Antarctica. Wegener added the northern supercontinent (called Laurasia after Asia and the Laurentian Mountains of the Precambrian shield of Canada) to Gondwanaland to

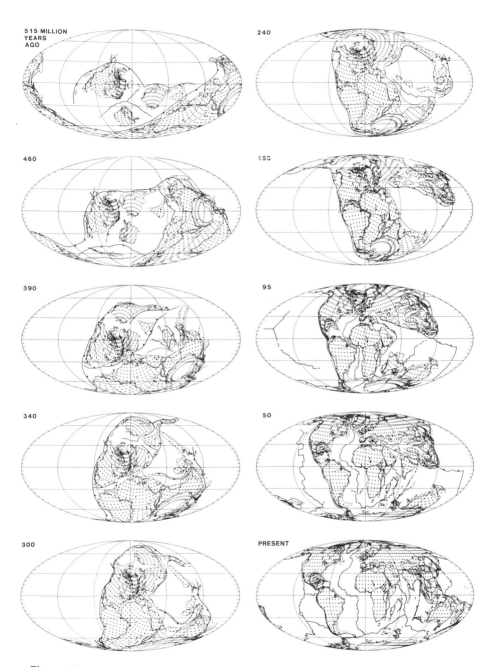

Figure 14
Reconstruction of drifting continents from 540 million years ago to the present. *(Paleomap Project, Christopher R. Scotese, University of Texas, Arlington)*

make Pangaea. The earlier Atlantic Ocean, present from 540 million to 300 million years ago, was called the Iapetus Sea. At 540 million years ago, just after metazoan life evolved, there was a strange, unrecognizable configuration of continents. Most every type of extant plant and animal organism on Earth—except perhaps some kinds of vascular plants and small, insignificant phyla—evolved between 540 million and 300 million years ago, before the gradual assembly of the supercontinent. Pangaea was complete by 240 million years ago. Between 240 million and 180 million years ago, nothing much happens tectonically except in the outer Pacific continents. The central and northern Atlantic Ocean begins to open by 120 million years ago, the south Atlantic by 60 million years ago. During the Cenozoic era, beginning 65 million years ago, Greenland opens up, and so does the north Atlantic off the coast of Scotland. We arrive at the current configuration when the Indian subcontinent drifts northward from Antarctica and rams up against the Asian continent to form the Himalayan Mountains.

In the last few years it has become possible to put together a geological map of the world (figure 15). The banded parts show relatively simple ocean basins dominated by the magnetic anomalies. An orderly pattern, symmetrical around the ridge, shows basalt covered by various layers of sediment. The continents, which preserve a record to almost 4 billion years ago, are more disorderly. The North American Cordillera from the Rocky Mountains to the Pacific Coast is an example of an exceedingly complex terrain: various slivers of continent were moved around by plate motions, docked at the main continental mass, and accreted to it.

When did plate tectonics begin? A few believe that these patterns characterize only the last 100 million or 200 million years of geologic history, but most place the inception much earlier. A great many geologists now agree that there is evidence of plate tectonics in the Vendian (late Precambrian). Whether plate tectonics was established as far back as the Archean eon, the period before 2.5 billion years ago, is subject to some debate among specialists. We grope for mechanisms responsible for the cooling of the early Earth. Did magnetic reversals operate in the same way in the ancient past as they have in the recent past? How big were the early continents? (We suspect they were much smaller than the continents today, but that's another story.)

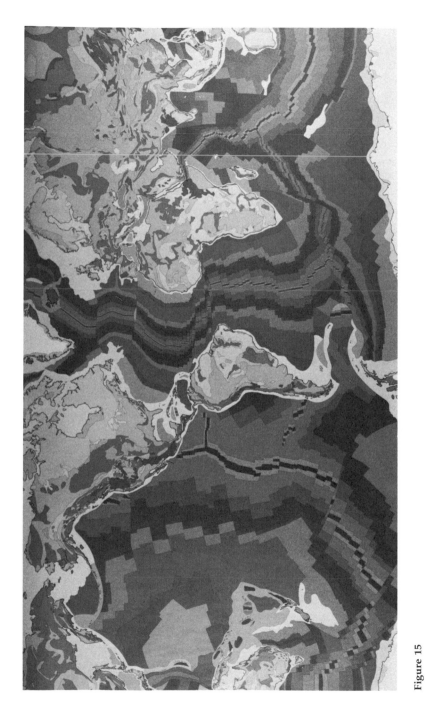

Figure 15
The bedrock geology of the world. Ocean basins are easily recognized by the parallel patterns of magnetic reversals. (*From R. L. Larson and W. C. Pitman, The Bedrock Geology of the World (Freeman, 1985). Copyright 1985 R. L. Larson and W. C. Pitman. Reprinted with permission.*)

Similar geophysical processes, such as convection currents, probably have been at work on the moon, on Mars, and maybe on other planets. However, the thickness of a lithosphere is very much a function of the temperature and thus of the heat flow from the interior of the planet. The moon has lost most of its interior heat. Although the moon's lithosphere may be as much as 400 km thick—too thick to allow any breakage and movement—there is probably little or no convection in its interior. Mars appears to be an intermediate stage. Mars, like the moon, was more active earlier in its history. It may still be intermittently active, or maybe it will become active if enough heat builds up. No evidence exists now for plate motions on Mars, and we don't know whether any convective motion occurs in the Martian mantle. Plate motions can only be inferred from surface manifestations of convection currents. Venus, we believe, has enough interior heat to maintain a sufficiently thin lithosphere so that plate movements and mountains may occur. Venus is blanketed by a dense atmosphere through which we can see only by means of sensitive instruments. There is evidence of volcanoes on Venus, but we still do not know enough about the surface of Venus to infer plate tectonics there. We expect the results of the Magellan mission to be illuminating.

Plate tectonics is an enormous milestone in our understanding of the history of the Earth. We now have an all-encompassing theory that explains the major features of our planet's surface, both the continents and the ocean basins. Further, the theory, based on fundamentally sound geophysical ideas of how the Earth loses heat, links the history of the atmosphere to that of the oceans, providing a mechanism for chemical exchange from the interior to the exterior of the Earth. Together, plate tectonics and continental drift theory are to twentieth-century geology what Newtonian mechanics was to pre-relativistic physics. The theory still lacks details of the mechanics. What dynamical forces pull the plates apart? Many think that lateral plate movement is driven less by "push" from mid-ocean ridges than by "pull" as plates descend into the mantle. Many questions need answering—but plate tectonic theory, constantly in revision, is very much alive. For those of us working on the mechanisms of the Earth's surface, and particularly on the interaction of those mechanisms with life, plate tectonics is the foundation on which future theories of all surface interactions on the Earth will be based.

Readings

Cox, A., ed. 1973. *Plate Tectonics and Geometric Reversals*. Freeman.

Menard, H. W. 1986. *The Ocean of Truth: A Personal History of Global Tectonics*. Princeton University Press.

Press, F., and R. Siever. 1986. *Earth*. Freeman.

Uyeda, S. 1978. *The New View of the Earth: Moving Continents and Moving Oceans*. Freeman.

Van Andel, T. H. 1985. *New View of an Old Planet: Continental Drift and the History of the Earth*. Cambridge University Press.

11

Chemical Signals from Plants and Phanerozoic Evolution

Tony Swain

As the late Tony Swain discusses in this chapter, secondary compounds are not just waste products of no further use to the organisms that produce them; many are, in fact, "ecological hormones." These compounds have ecological roles as semiochemicals (chemicals that mediate interactions between organisms in the environment)—that is, they are chemical signals. Two types of semiochemicals are recognized: pheromones (compounds produced by and targeted toward members of the same species) and allelochemicals (compounds produced by members of one species and evoking responses in members of different species).

In 1986, Robert Buchsbaum (a former student of Swain's who is now the Massachusetts Audubon Society's Coastal Ecologist) addressed questions raised by Swain's lecture. The questions and Dr. Buchsbaum's answers follow Swain's text.

It has long been recognized that plants and animals profoundly affect the characteristics of one another during the course of their evolution. We have only to consider the case of pollination of flowering plants to see the truth of this statement. Both flowers and their pollinators, whether they are insects, birds, or bats, have developed specialized adaptations to promote success of the operation. Flowers attracting animals with bright colors ensure that pollinators enter their tissue by rewarding the animals with pollen and nectar. This form of signaling between plants and animals is really the essence of biochemical coevolution.

Chemical signals between organisms are largely due to compounds that are not part of primary metabolism. Primary metabolites are absolute requirements for existence and reproduction, such as carbohydrates, nucleic acids and their components (nucleotides, nucleosides, etc.), proteins, amino acids, and lipids. Chemical signals emitted by organisms have been described in the literature as "secondary compounds" or "secondary metabolites," presumably to indicate that they are not as important as primary metabolites. Organic chemists have traditionally referred to these compounds as "natural products" to distinguish them from synthetic compounds made in laboratories. From the ecological and evolutionary point of view, the term *semiochemicals* (from the Greek word *semio*, meaning "signal" or "sign") is more appropriate than these other terms, since it emphasizes their role as chemical signals between organisms. Semiochemicals can serve many ecological roles, such as prey or mate attractants, feeding deterrents, or warnings and deterrents to potential predators or competitors. Semiochemicals are produced by an organism; then either they are emitted into the environment or they remain within the tissues of the producer and eventually evoke a response in another organism.

In animals, neuroreceptors receive chemical signals, and these neuroreceptors send signals to the brain. In most mammals the nose and the mouth are the only organs that receive both smells and tastes, but in insects we find receptors very similar to taste receptors on the feet and other parts of the body. Many insects have feathery protrusions from the tops of their heads which, like our noses, are able to receive and amplify chemical signals of volatile compounds. The emission and reception of these compounds from one insect to another helps regulate their mating, aggregation, defense, and other behaviors.

In this chapter we examine only those semiochemicals produced by plants. We ask how some may have changed during the course of evolution in response to the insects that coevolved with the plants. Among extant organisms, we find that land plants, by and large, produce the largest number of semiochemicals. Plant semiochemicals can be assigned to several different classes on the basis of their chemical structures. We consider here one or two examples from the three major classes: phenols, terpenes, and alkaloids (table 1). The nearly 11,000 known structures are only the tip of the iceberg; the number of compounds present in nature is probably 10 or even 20 times greater.

Table 1
Major classes of semiochemicals.

	Approximate number of structures known
Phenols	
flavonoids[a]	⎫
tannins	⎬ 1,000
	⎭
lignins	not well characterized (structurally complex)
Terpenes[b]	
monoterpenes	1,000
sesquiterpenes	600
diterpenes	1,000
triterpenes	800
tetraterpenes	350
polyterpenes	?
Alkaloids	6,000

a. See figure 1A.
b. See figure 1C.

Phenols as Semiochemicals

A typical phenolic compound, a flavonoid compound called an anthocyanin, is shown in figure 1A. One common anthocyanin, cyanidin,
is responsible for the red color in roses and in the skins of winesap
apples and other red fruits, such as cherries. The bright red of this
anthocyanin is important for flower pollinators that see the color, and,
in the case of seed dispersal, for animals that see the red color of
fruits such as apples, plums, and cherries. Related anthocyanins are
responsible for scarlet, pink, violet, blue, and even black pigments in
flowers.

The flavonoids are phenols because they have aromatic or benzene
rings, to which hydroxyl (—OH) groups are attached. Flavonoids are
probably present in all of the flowering plants, the angiosperms, but
they are mostly absent from seedless vascular plants. Some are present
in the green algal group, the Characeae, believed to be the closest
relatives to plants. They are certainly not present in either bacteria or
fungi.

Basic Anthocyanin Structure – A Flavonoid

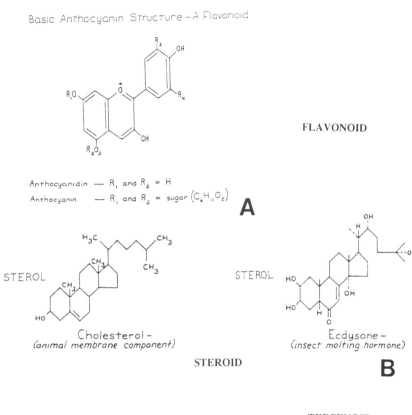

FLAVONOID

Anthocyanidin — R_1 and R_2 = H
Anthocyanin — R_1 and R_2 = sugar $(C_6H_{11}O_5)$

A

STEROL

Cholesterol –
(animal membrane component)

STEROL

Ecdysone –
(insect molting hormone)

STEROID

B

TERPENOID

SESQUITERPENE

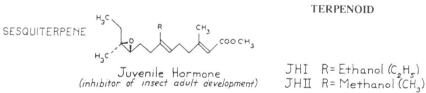

Juvenile Hormone
(inhibitor of insect adult development)

JHI R = Ethanol (C_2H_5)
JHII R = Methanol (CH_3)

C

Figure 1
A: An anthocyanin, a type of flavonoid that is an important flower color constituent. An anthocyanidin possesses hydrogen at R_1 and R_2; an anthocyanin has sugars substituted at these points. B: The triterpenoid sterols, cholesterol and ecdysone. Ecdysone, chemically very similar to cholesterol, is an insect hormone involved in molting. C: The sesquiterpenoid insect juvenile hormone. *(drawing by Sheila Manion-Artz)*

Tannins and lignins are two other groups of phenolic compounds that are essential in many interactions between vascular plants, herbivores, and plant pathogens. The condensed tannins are polymeric molecules, usually containing five to ten flavonoid units joined together (figure 2). Their molecular weights range from 1500 to 3000 daltons. A second group of tannins, the hydrolyzable tannins, have a "core" sugar molecule to which various simple phenol molecules are attached. Tannins act as deterrents in nature by forming hydrogen bonds between their phenolic hydroxyl groups and the peptide links and other reactive groups of proteins or the hydroxyl groups of polysaccharides. All phenols bind with proteins to some extent; however, the particular effectiveness of tannins as defense compounds in plants relates to their ability to form large insoluble complexes with digestive enzymes of animals, plant proteins, and carbohydrates. This interferes with the digestion and assimilation of nutrients by animals.

Humans, like many animals, avoid plants high in tannins if given a choice. Most of the plants eaten by humans have been selected over many centuries by agriculturists and food producers to be low in distasteful substances such as tannins and other deterrents. If you want to experience for yourself what tannins taste like, bite through an unripe persimmon, grape seeds, or a banana peel. The dry, puckery taste quality of tannins is known as *astringency*.

Tannins are particularly characteristic of woody plants. Although they may occur in all parts of the plant, they are typically most abundant in woody tissues, where they not only deter herbivores but also deter diseases and delay decomposition by fungi. Plants that evolved early in geologic time, such as mosses, liverworts, and lycopods, lack tannins. The first group of land plants to contain tannins, the ferns, evolved in the early Carboniferous or perhaps the late Devonian period, around 360 million years ago. Modifications in the biosynthetic pathways of flavonoids in this group allowed the evolution of tannins in ferns, gymnosperms, and angiosperms. Hydrolyzable tannins are of even more recent origin, since they occur only in dicots.

Lignins are well-known phenols because they are part and parcel of the cell wall of woody plants. Lignins are what we might call the essence of woodiness. They act as plastic glues, locking in the long-chain cellulose molecules that act as reinforcing bars of the cell walls.

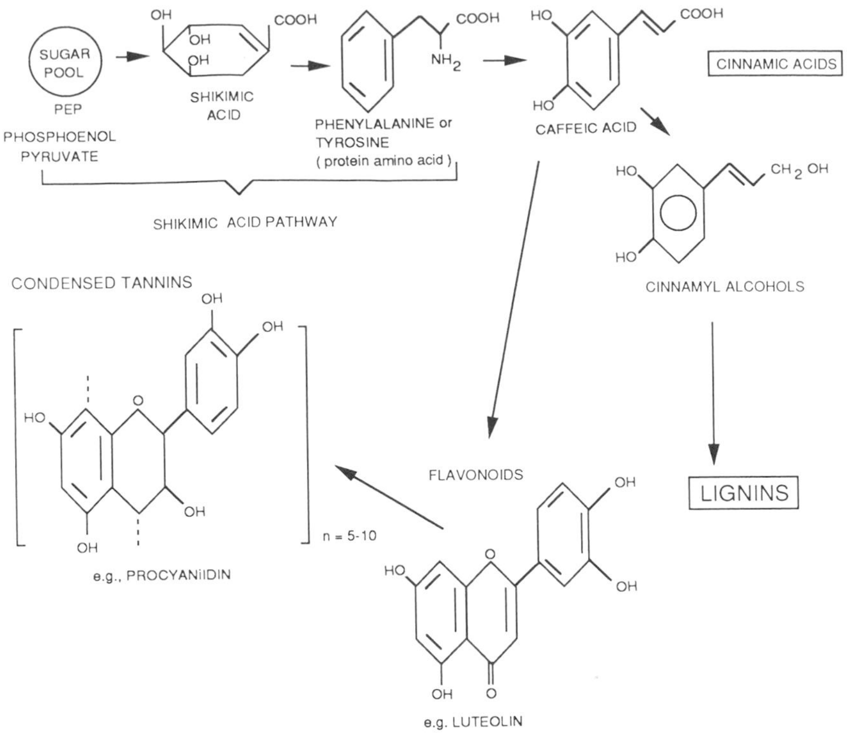

Figure 2
Biosynthetic relationships among flavonoids, tannins, and lignins. Phenylalanine and tyrosine, protein amino acids derived from phosphophenyl pyruvate via the shikimic acid pathway, are the precursors of cinnamic acids. Cinnamic acids are used to make flavonoids and lignins. Condensed tannins are derived from flavonoids. *(drawing by Sheila Manion-Artz)*

This makes cell walls tough and woody, and therefore not desirable as food.

The lignins evolved somewhat earlier than the tannins. We find lignin in all plants that contain transporting vessels, including the modern representatives of the very earliest plants visible in the fossil record. We find no lignins in protists or in bacteria, as we might expect. They do occur, however, in *Psilotum*, a present-day remnant of the Rhyniophyta (figure 3). They also occur in lycopods, in horsetails, in ferns, and, of course, in the gymnosperms and angiosperms.

At the end of the Devonian period, around 360 million years ago, there were among the lycopods some trees 100–120 feet tall. It seems likely that lignins were necessary not only to ensure upward growth,

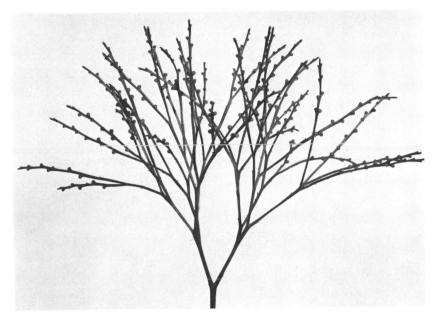

Figure 3
Psilotum, a member of the Rhyniophyta, is a living relative of the earliest vascular plants.
(William Ormerod)

but also to ensure the strength of the vessels that transported water up from the roots.

The evolution of these phenolic compounds and their biosynthesis is probably very ancient. They all arise from the aromatic protein amino acids phenylalanine and tyrosine. These universal metabolites are found in bacteria; we assume they appeared in the early Archean eon or before. The aromatic amino acids are formed from the sugar pool by a relatively straightforward route which includes the simple alicyclic (not aromatic) compound shikimic acid.

The biosynthetic and evolutionary relationships among flavonoids, lignins, and tannins are depicted in figure 2. Flavonoids are present in primitive land plants (*Psilotum*, lycopods, etc.) and, as mentioned above, in one group of green algae: the Characeae of lakes and slow-moving streams. These compounds may initially have developed as light screens to protect early land plants against photodestruction by ultraviolet light of nucleic acids and some coenzymes. The complexity of these compounds increases with evolution; the most recently appearing plants have the most complex flavonoids. Lignins first appear in early land plants, imparting structural rigidity in air. Tannins first appear as flavonoid derivatives in ferns.

The evolution of phenols as defense compounds is related to the movement of plants to land. Plants first coming onto land were probably accompanied by some herbivorous insects. The amphibians did not evolve until much later, followed even later by reptiles and mammals, many of which are voracious plant eaters. Nevertheless, the early insects were probably important herbivores of land plants, eating spores or sucking plant juices. Certainly, a large number of early insects were detritivores, chewing up plant remains along with the bacteria and fungi that accompanied plants onto land.

These herbivore-deterrent compounds, especially tannins and lignins, are obviously important in relation to the development of insects. As lignin enabled plants to evolve into taller forms, insects evolved wings; therefore, height did not provide plants with complete protection from insects. Although insect wings may have evolved at first to regulate body temperature, they allowed insects to fly higher and to exert selection pressure on plants such that even taller plants evolved. Similarly, the development of tannins in leaves and stems, but not in spores, prevented herbivores from chewing on those tissues in plants. We may suppose that lignins and tannins had a marked effect in that a large number of plant remains were not eaten during the Carboniferous period, leading to the extensive coal measures that we find from that period.

Isoprene Units and Terpenoid Formation

The terpenoids are another class of compounds that certainly had enormous effects on the coevolution of plants and animals. Turpentine, produced from pine trees, is an oily liquid formed from a number of terpenoid compounds present in many gymnosperms. The terpenoids are formed from a spiky C-5 unit called isoprene (figure 4). The basic unit that condenses to form the polymers is not isoprene itself but a pyrophosphate derivative called isopentenyl pyrophosphate (IPP) and its isomer dimethylallyl pyrophosphate (DMAPP). The synthesis of isopentenyl pyrophosphate, which comes from the universal 2-carbon metabolite acetate, is shown in figure 4. IPP condenses in a head-to-tail fashion (figure 5) to give dimers, trimers, and so on containing 5, 10,

A. Isoprene (C_5H_8)

$$CH_3$$
$$\underset{H_2C}{\overset{CH_3}{\bigg/}} C \underset{CH}{\overset{CH_2}{\bigg\backslash}} \equiv \text{tail} \qquad \text{head}$$

B. Isopentenyl pyrophosphate

IPP   Terminology

$$\underset{H_2C}{\overset{CH_3}{\bigg/}} C \underset{CH}{\overset{CH_2}{\bigg\backslash}} O - \overset{O}{\underset{O^-}{\overset{||}{P}}} - O - \overset{O}{\underset{O^-}{\overset{||}{P}}} - O^- \equiv \text{tail} \quad \text{head} \quad \text{OPP}$$

acetate residues

$$CH_3 - \overset{O}{\overset{||}{C}} - CH_2 - \overset{O}{\overset{||}{C}} - SCoA \; + \; CH_3\overset{O}{\overset{||}{C}} - CoA \; + \; H_2O \longrightarrow \begin{matrix} COO^- \\ | \\ CH_2 \\ | \\ HO - C - CH_3 \\ | \\ CH_2 \\ | \\ C - S - CoA \\ || \\ O \end{matrix} \longrightarrow$$

Acetoacetyl CoA   *Acetyl CoA*

Synthesis

3-Hydroxy-3-methyl-glutaryl CoA

$$\longrightarrow \begin{matrix} COO- \\ | \\ CH_2 \\ | \\ HO - C - CH_3 \\ | \\ CH_2 \\ | \\ CH_2OH \end{matrix} \quad \overset{3\,ATP \quad 3\,ADP}{\underset{H_2O + CO_2 + P_i}{\longrightarrow}} \quad \begin{matrix} CH_2 \\ || \\ C - CH_3 \\ | \\ CH_2 \quad O \quad O \\ | \quad || \quad || \\ CH_2O - P - O - P - O^- \\ | \quad | \\ O^- \quad O^- \end{matrix}$$

Mevalonate   *Isopentenyl pyrophosphate*

IPP

C.

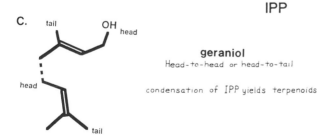

geraniol
Head-to-head or head-to-tail

condensation of IPP yields terpenoids

Figure 4
A: Isoprene and isopentenyl pyrophosphate. Terpenoids are composed of units of the five-carbon compound isoprene. The pyrophosphate derivative of isoprene, isopentenyl pyrophosphate, condenses to form the terpenoids. B: Isopentenyl pyrophosphate is synthesized from acetate in the form of acetoacetyl CoA and acetyl CoA. C: Geraniol is a typical monoterpene, formed by the head-to-tail condensation (broken line) of two isoprene units. *(drawing by Sheila Manion-Artz)*

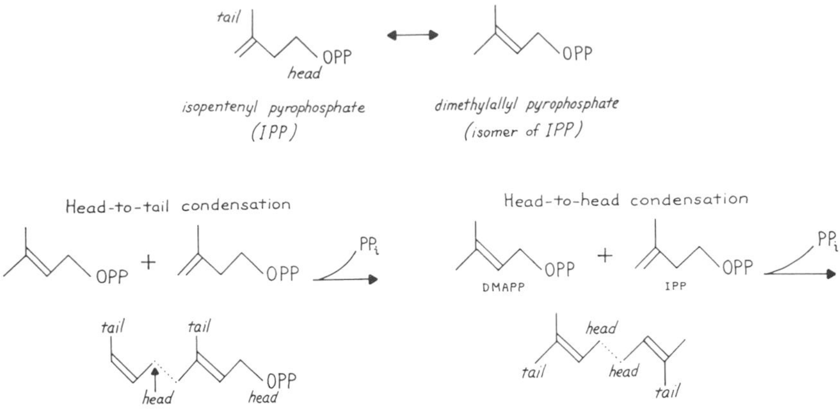

Figure 5
Head-to-tail or head-to-head condensation of isopentenyl pyrophosphate (IPP) and its isomer, dimethylallyl pyrophosphate (DMAPP), yields terpenoids. *(drawing by Sheila Manion-Artz)*

15, 20, or more carbons. The biosynthetic pathways of the isoprenoids are obviously very ancient, because such compounds occur in many of the quinones that we find even in anaerobic bacteria, used in the transport of electrons from one side of a membrane to another. Bacteria also use isoprenoid compounds, usually containing 50–80 carbon atoms (10–16 isoprene units) to transport sugars from where they are manufactured within the cell to the cell wall for incorporation into the various units there. Another group of terpenoids—the carotenoids (figure 6), required in all photosynthetic systems—are present in photosynthetic bacteria and all photosynthetic eukaryotes. Bacteria chlorophylls also contain a side chain onto which three or four isoprene units are attached. These examples indicate that the pathway to produce three or four isoprene units is a very ancient one.

Terpenoids as Semiochemicals

I have mentioned terpenoid quinones, and terpenoids that transfer sugars in bacteria, and the carotenoids. Making an enormous jump in time, let us now shift to the flowering plants (which evolved only 120 million years ago) and the gymnosperms (which are a little older, having evolved about 220 million years ago). Both groups of plants

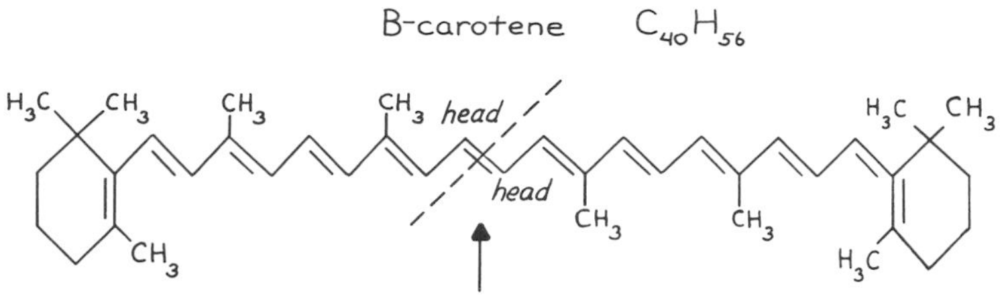

B-carotene $C_{40}H_{56}$

Figure 6

Carotenoids, C_{40} compounds, contain a total of eight isoprene units. They are composed of two groups of four isoprene units condensed head to tail, which are then joined (at the arrow) by head-to-head condensation. *(drawing by Sheila Manion-Artz)*

contain a vast number of semiochemical terpenoid compounds which act as signals.

Semiochemical terpenoids tend to be simple, containing only two or three isoprene units (that is, 10 or 15 carbon atoms condensed head to tail). They are called monoterpenes if they have 10 carbon atoms, sesquiterpenes if they have 15 carbon atoms (figure 7). Four isoprene units can condense to form diterpenes. Monoterpenes, sesquiterpenes, and diterpenes can be semiochemicals in seed plants, but they are rare or virtually absent in plants that lack seeds. They are absent or uncommon in fungi, animals, and protists. They are completely absent from the bacteria. So, we have a paradox here. The early method of biosynthesis, which could produce large numbers of isoprene units joined together, has been superseded by one which gives simpler types. We can think of this in relation to the biosynthesis of fatty acids. The majority of fatty acids we find in membranes or in storage cells, in plants and animals, contain 16 or 18 carbon atoms, and they are made by the continued addition of two-carbon acetate units to produce the final 16 or 18 carbon atoms. Fatty acids containing fewer carbons are much rarer, pointing to the same paradox: low-carbon-number terpenoids are nearly absent in the predecessors to plants. Their evolution was presumably due to some challenge from animals that led plants to vary the control of their terpenoid biosynthetic pathway so that they produce these low-molecular-weight compounds that act as semiochemicals.

How do monoterpenes, sesquiterpenes, and diterpenes act as semiochemicals? The low-molecular-weight monoterpenes are mainly

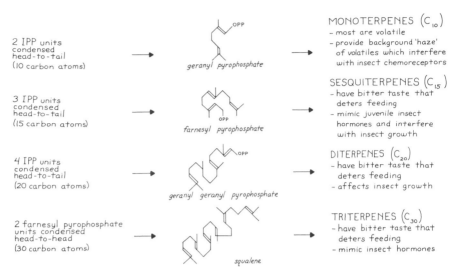

Figure 7
Terpenoid compounds: components, representative structures, and properties of monoterpenes, sesquiterpenes, diterpenes, and triterpenes. *(drawing by Sheila Manion-Artz)*

volatile. They account for the piney smell in conifer forests, and for most of the scents of angiosperm flowers. They provide a background of scents that can be attractive to pollinators, but also can act as a kind of hazing mechanism, which prevents insect volatile scents from being properly evaluated by the receivers.

The sesquiterpenes are far less volatile. Their importance like that of many of the alkaloids, lies in deterring feeding by their bitter taste. This applies especially to those sesquiterpenoids that have one or more hydroxyl or other oxygen residues substituted in the molecule, giving rise to internal lactones. The very bitter sesquiterpenoid lactones deter feeding by insects and mammals. Some also affect the growth and development of insects, and perhaps the development of other animals as well. Diterpenoids, which also affect animal growth, act as feeding deterrents; many of them, especially those that contain oxygen, are bitter.

Sesquiterpenoids very similar in structure to a hormone of insects occur in some gymnosperms. "Juvenile hormone," when released by an insect during molting, retains the insect in an immature phase, preventing it from maturing to an adult. In butterflies and moths, the caterpillar grows in size, going through its various molts, but juvenile hormone keeps the insect as a caterpillar until the very last molt, where it develops not into a larger caterpillar but into a pupa. In this last molt the insect juvenile hormone is not released. Some plant analogues of insect juvenile hormone have an enormous effect on some insects: they ensure that these insects become very giant juveniles and never mature into adults.

One more class of terpenoids, extremely important to us all, is the triterpenoids. Like the carotenoids, triterpenoids are formed by a final head-to-head joining, but in this case isoprenes (two C-15 units) join to give a final 30-carbon compound that then forms rings. The importance of the triterpenoids is that they also yield sterols. Sterols are important in membranes of all eukaryotes. In animals they also act inside the body as internal messengers which we call hormones. (The structure of cholesterol is shown as B in figure 1.)

A large number of triterpenes also are feeding deterrents because of their bitter taste. These compounds are present in many plants that we would find bitter. Cucumbers, for example, would be extremely distasteful were it not for selective breeding for low concentration of triterpenes. Ecdysones are among the most important of the triterpenoids. These are again insect hormones affecting the overall molting procedure; they are produced from sterols by small chemical changes in the side chain (figure 1, B and C). Unlike juvenile hormone, which merely keeps the insect juvenile, ecdysones act as hormones that ensure molting will take place. Ecdysones are very similar to the common sterol cholesterol, which is present in our bodies and those of all mammals. Many ferns, and some gymnosperms, produce very large amounts of terpenoids that mimic insect hormones. It is suspected that until insects evolved ways to properly detoxify and cope with them, these sterols were extremely troublesome. They certainly do not entirely inhibit the development of modern insects, but we assume that during the course of evolution such sterols affected some kinds of insects drastically. Some may still do so.

The carotenoids are a ubiquitous class of terpenoids that contain eight isoprene units, consisting of two groups of four head-to-tail isoprene units joined together head to head (figure 6). The carotenoids are present not only in photosynthetic organisms but also in heterotrophic bacteria, as well as in protists, fungi, and animals. A large number of animals sequester carotenoids from plants and use them for their own coloring. The color of lobsters, for example, is due to a carotenoid-protein complex. The use of different terpenoids must have varied during the course of evolution. Carotenoids provide a good example of how this may have happened. Carotenoids are present in all photosynthetic bacteria; they are essential to the light-harvesting system. However, in flowering plants, carotenoids have taken their part, as with the flavonoids, in petal pigments of flowers and in the skin color of fruits. The bright yellow of daffodils and the red of tomatoes are good examples. Evolutionary processes seem to use ancient chemical pathways to produce signals which lead to plant and animal coevolution.

Alkaloids from Amino Acids

Most of us are aware of the last group of semiochemicals to be discussed here, the alkaloids, since coffee contains caffeine. Many of us are familiar too with poisonous alkaloids such as strychnine, which are still used to rid houses of mice, rats, and other rodents. And many of us are acquainted with the dangers of the alkaloids morphine, heroin, and cocaine.

Most alkaloids are plant products, although a few are produced by animals. For example, the skins of certain frogs and toads in South America contain alkaloids. The main characteristic of alkaloids is that they all contain nitrogen and, in solution, they all act to take up protons (i.e., as "bases" in the sense of raising pH to create alkaline solutions). Usually the nitrogen is present within a six-member ring. These heterocyclic rings, as they are called, often contain five carbon atoms and a nitrogen atom. Another defining characteristic of alkaloids is that they are highly physiologically active on animals. Some, like strychnine, are poisonous; others, like heroin, have deleterious physical or psychological effects. Although people may at first consider the effects of morphine and heroin euphoric and pleasant, heavy use of these substances leads to dependence and degradation and even death. Alkaloids have the same kind of effects on many animals, acting as deterrents or attractants as the case may be.

Unlike the other two classes of compounds, terpenoids and phenolics, almost 99 percent of all known alkaloids occur only in the flowering plants; therefore, they have evolved only during the last 120 million years or so. Some alkaloids are found in fungi, some in lycopods, and one or two in gymnosperms. The alkaloids contained in lycopods are based on lysine. The pathway from lysine to some alkaloids is relatively early; even the general route to the protein amino acid lysine may include some cyclic compounds. But, by and large, we are surprised that not until the mammals evolved did plants scramble to produce the large number of new biosynthetic pathways that enabled them to synthesize the extraordinary variety of alkaloids that deter insects and mammals today. Certainly no alkaloids have been reported to be synthesized by protoctists or prokaryotes. Let us consider exactly how alkaloids are formed in the flowering plants.

I am limiting my discussion here to alkaloids formed from amino acids. They fall into two classes: those formed from the aromatic amino acids phenylalanine, tyrosine, and tryptophan and those formed from the diaminoaliphatic amino acids lysine and ornithine.

Not until the evolution of the flowering plants were these alkaloids produced in sufficient quantities to act as feeding deterrents, yet their mode of biosynthesis from primary metabolites is relatively simple (figure 8). Phenylalanine is decarboxylated, and the resulting amine, phenylethylamine, is condensed with an aldehyde to close the ring and create the various alkaloids (figures 8 and 9). The same is generally true of the nonaromatic amino acids, such as lysine, except here the amino acid is not only decarboxylated; it also loses the alpha-amino group, and its terminal amino group forms a ring which closes to produce alkaloids such as cocaine and nicotine (figures 8 and 10).

The most physiologically active of the alkaloids are those derived from aromatic amino acids. These include benzoisoquinoline alkaloids (figure 9), which lead to the formation of compounds such as morphine and which are known not only in flowering plants but also in certain fungi.

The alkaloids, almost without exception, have a bitter taste. They may be extremely bitter or only slightly so. Their taste gives us and other mammals a clue to their occurrence. Excessive bitterness in food tends to repulse most organisms. I believe "bitterness" is the most primitive of all the taste sensations. Alkaloids synthesized from the aromatic amino acids are often the most bitter and usually the most toxic.

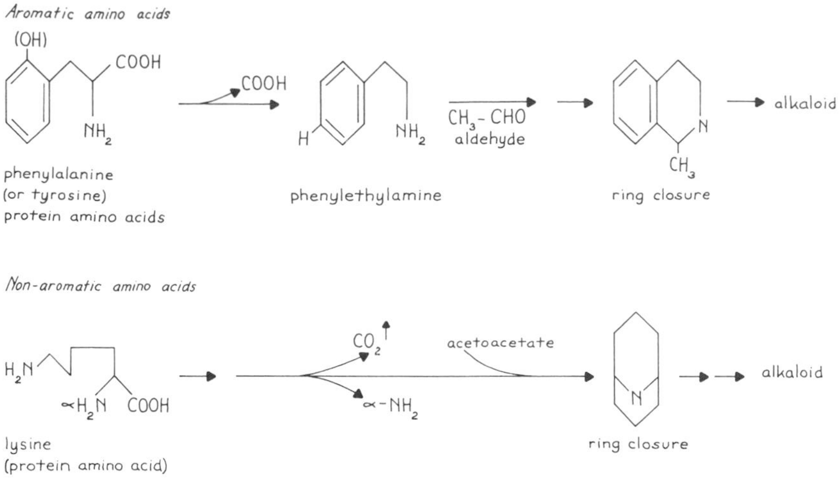

Aromatic amino acids

phenylalanine
(or tyrosine)
protein amino acids

phenylethylamine

aldehyde

ring closure

alkaloid

Non-aromatic amino acids

lysine
(protein amino acid)

acetoacetate

ring closure

alkaloid

Figure 8
Biosynthesis of alkaloids from amino acids. Aromatic amino acids are decarboxylated and condensed with an aldehyde to form the closed nitrogen-containing ring structure of various alkaloids. In nonaromatic amino acids, the -amino group is lost during decarboxylation and condensation with acetoacetate. The terminal amino group becomes incorporated into the closed ring of the resulting alkaloid. *(drawing by Sheila Manion-Artz)*

Returning to the biochemical coevolution of plants and animals, we can start looking at land plants and land animals from the early Devonian period to the present. Plants, early in their evolutionary history, had to cope not only with insects and other animals but also with fungal, bacterial, and viral pathogens.

The Role of Semiochemicals in the Evolution of Land Plants

We do not know if the first land plants had to cope with attack by nematodes, earthworms, or the land snails present as soil fauna today, because the fossil record on this issue is sparse. The first accounts of land snails do not occur until about 330 million years ago. Nevertheless, plants have thrived on land since the Devonian period. They must have had bacterial, fungal, and animal helpers that ensured production of soil. Detritus was enzymatically chewed up, and particles of sediment were attracted and collected. Such coevolution suggests not only

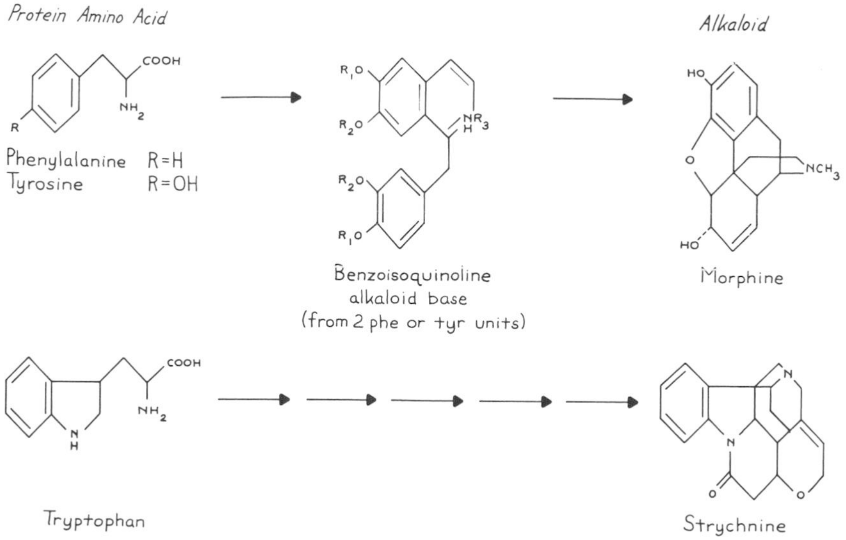

Protein Amino Acid   *Alkaloid*

Phenylalanine R=H
Tyrosine   R=OH

Benzoisoquinoline
alkaloid base
(from 2 phe or tyr units)

Morphine

Tryptophan

Strychnine

Figure 9
Alkaloids derived from aromatic amino acids. The synthesis of morphine from pheny-
lalanine and tyrosine involves the formation of benzoisoquinoline alkaloid intermediates.
Strychnine is made from tryptophan in a many-step metabolic pathway. *(drawing by
Sheila Manion-Artz)*

fierce battles between plants and animals but also strong cooperative
interactions. The evolution of lignins and then tannins during the late
Devonian/early Carboniferous period provided a way for plants to
avoid the fiercer insect herbivores that had evolved by then. Perhaps
at this time plants developed insect hormones, such as juvenile hormone and the ecdysones responsible for molting. During the Carboniferous period, when a great diversity of large land plants thrived and
prospered, detritus was not readily decomposed. Perhaps some plant
defenses were far too strong for the animals, mainly insects, living at
that time.

The first of the land vertebrates, the amphibians, probably did not
have much of an effect on plants. Most modern amphibians are carnivores, eating insects; perhaps the early ones were also. Not until the
reptiles evolved, toward the end of the Paleozoic Era (about 290
million years ago), moved away from the river and lakeside habitats
of the amphibians, and became the dominant animals were plants
forced to cope with more voracious herbivores. Probably during that

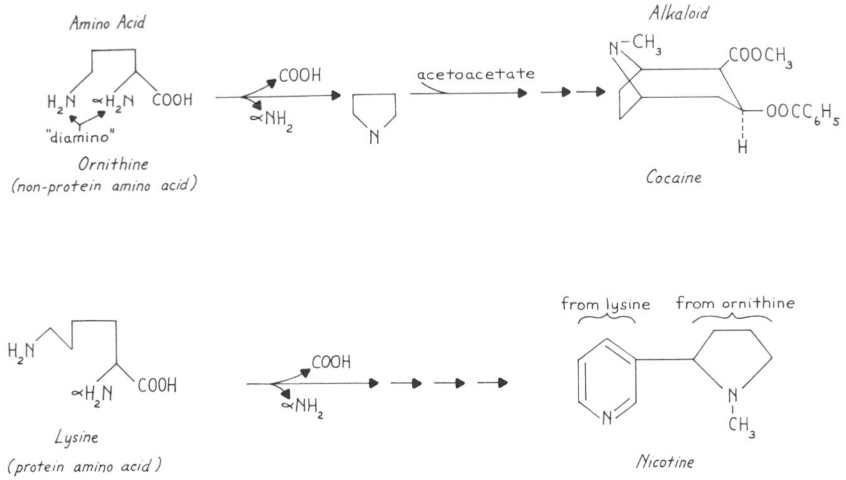

Figure 10
Alkaloids formed from diaminoaliphatic amino acids. Cocaine and nicotine are formed from ornithine and lysine, respectively. *(drawing by Sheila Manion-Artz)*

time the gymnosperms, the ferns, and a number of other plant orders that have since died out developed new types of deterrents. The various terpenoids we find in present-day gymnosperms (monoterpenoids, sesquiterpenoids, diterpenoids, and triterpenoids) apparently evolved in response to these challenges.

When the angiosperms finally evolved, about 120 million years ago, they radiated extensively, becoming dominant plants very quickly—within 60 million years. Concurrently, a large number of new types of compounds evolved, including chemical signals produced via old biosynthetic pathways. Colored compounds and a host of alkaloids and terpenoids developed and were used in different ways. Plants had long been using lignin and tannins as deterrents; animals and other organisms still have not devised suitable ways to overcome them. (Of course, humans, beavers, and packrats have exploited the elastic lignin for various uses, including firewood, homes, and buildings.)

We can look forward to knowing more and more about the biochemical coevolution of plants and animals as more work is devoted not just to the structures of natural products so beloved by the organic chemists but also to the way in which these compounds act as semiochemicals and the way in which animals have devised detoxification and other mechanisms to overcome them.

Questions and Answers

Robert Buchsbaum

What classes of volatile chemicals can our noses detect?

Our noses can detect monoterpenes, simple phenols, amines, and sulfur-based compounds. Monoterpenes include many familiar odors of flowers and flavorings: e.g., lemon, lime, and menthol. Simple phenols include compounds such as cinnamon and vanilla. The amines are responsible for the odors of cooking and rotting meat. For example, the amines putrescine and cadaverine are named for their smell. The smells of garlic and onions, as well as the skunky odors that are found throughout nature, are due to sulfur compounds. We do detect other classes of compounds, but these are the major ones.

Are animal receptors sensitive to specific plant chemicals, or is the sensitivity generalized?

In some cases, particularly with invertebrates, there seem to be specific receptors; for example, lobsters have a specific receptor for glutamine. In vertebrates, it seems that the receptor is fairly generalized. For more information, see Visser 1983.

Plants seem to show more evolutionary change in their chemistry than animals; furthermore, plants apparently evolved chemical defenses against animals. Are these generalizations accurate?

Plants display more chemical evolution because they don't "behave"; they can't run away the way animals do. The major response by plants to attack by animal herbivory is the evolution of defensive chemicals. Plants, of course, respond or evolve to many different factors (not just to herbivores), including light, nitrates, phosphates, and other nutrients, and to the attraction of pollinators.

How do animals generally adapt to evolutionary changes in plants?

Animals tend to change their behavior and, with time, evolve different behavioral mechanisms. They may simply avoid eating toxic plants, or they may avoid toxic parts of plants, as aphids do by feeding on phloem sap.

In addition (and this is fairly well studied), a variety of enzymes in animal guts, such as mixed-function oxidases, act to detoxify secondary compounds; they are very general in their activity. Animals also have an enzyme called rotenase, which detoxifies cyanide, a very toxic semiochemical in a number of plants. The addition of a thiosulfate group to cyanide, by this enzyme, yields thiocyanate, a compound much less toxic than cyanide.

How does hydrogen bonding of tannins to plant proteins and polysaccharides disrupt the ability of herbivores to digest the plant? Does the hydrogen bonding itself make them indigestible or bitter tasting? Can any animals or bacteria metabolize tannins?

According to current views, and there is still a lot of controversy on this issue, hydrogen bonding of tannins to proteins and polysaccharides occurs because the tannins are able to cross-link several protein molecules, creating large insoluble complexes. Chemical cross-linking occurs readily, but we don't know if it is the actual mechanism of their deterrence. Hydrogen bonding of tannins to proteins may make them indigestible, but they may also be repellent because of the astringent taste.

We assume that bacteria can metabolize tannins, and that is the way tannins are eventually broken down in the ecosystem. However, some insects deal with tannins by binding them within the gut to a membrane or lining, which is eventually egested.

What is the definition of an astringent? Are most astringents condensed tannins?

An astringent, in medicine, is something that constricts tissue, such as a substance that causes a wound to close. In the pharmacological sense, astringents such as alum are not necessarily condensed tannins. In ecological chemistry, astringency refers to the cross-linking of proteins, which gives a dry, puckery feeling in the mouth.

Are there animals that do not taste tannins as "bitter"?

We assume that other animals taste the tannins as what we call "astringent," just as humans do, but it's impossible to say with any assurance that these compounds cause other mammals or insects to experience the same taste we do.

In discussing terpenoids, [Professor Swain] refer[s] to head-to-tail and head-to-head condensation. Please explain these condensations and their significance.

Terpenoids consist of "spiky" five-carbon building blocks called isoprene units. The "head" and the "tail" are the two ends of the isoprene unit. The tail is at the spiky end of the molecule.

In the monoterpene geraniol (figure 4), a head-to-tail condensation of two isoprene units yields the ten-carbon compound. A generalized head-to-tail condensation is also illustrated on the left side of figure 5.

Smaller terpenoids (monoterpenes, sesquiterpenes, and diterpenes) are composed entirely of isoprene units (two, three, and four units, respectively) joined head to tail.

A head-to-head condensation is shown on the right in figure 5. For steroids and other triterpenes, two sesquiterpenes are joined together head to head to form a compound composed of six isoprene units. In the synthesis of carotenoids, two diterpenes are joined together head to head to form a 40-carbon compound (figure 6). The basic units of the carotenoids and other C_{40} terpenoids are isoprene units put together head to tail, but the final step in their synthesis involves one head-to-head condensation of C_{20} units.

Is there any evidence that monoterpenes prevent insect reproductive success?

Geraniol is a typical monoterpene, a terpene with ten carbon atoms (figure 4). Notice the "spiky" structure. Geraniol is an alcohol with a hydroxyl group. Monoterpenes are often feeding deterrents, or occasionally feeding attractants, so their effect on insect reproduction is indirect. Because the gymnosperms contain vast mixtures of monoterpenes, it is hard to determine the effects of a single monoterpene on insect reproductive success.

How do sesquiterpenes and diterpenes affect insect growth and development?

A monoterpene is made of ten carbon atoms, or two C-5 units, the basic isoprenoid building blocks of terpenoids. A sesquiterpene is composed of fifteen carbon atoms, and a diterpene contains twenty. Insect juvenile hormones are sesquiterpenes, and a number of plants (particularly the gymnosperms) produce mimics of juvenile hormones.

Sesquiterpene lactones comprise a well-known group of very bitter toxic compounds which occur primarily in sunflowers and other composites. Diterpenes are toxic or defense compounds in rhododendrons, *Kalmia* (mountain laurel), *Solidago* (goldenrod), and some chenopods (members of the spinach family).

What functions do the carotenoids have in plants? What is their function in bacteria? Are carotenoids limited to photosynthetic organisms?

Carotenoids are accessory pigments in all oxygenic photosynthesis, absorbing light in the parts of the spectrum where chlorophyll *a* does not. For example, the peaks of absorbance of carotenoids are 460 nm and 500 nm, whereas those for chlorophyll *a* are around 440 nm and 670 nm. Their conjugated structure, which includes a combination of single and double bonds, also protects plants from photo-oxidation. These double bonds can absorb ultraviolet light, and presumably work as UV screens. Present in all photosynthetic organisms (even bacteria), carotenoids are responsible for the red, orange, and yellow colors in some flowers and fruits (for example, tomatoes, *Calendula*, and lilies). Most of the yellows in flowers like *Rudbeckia*, the Black-Eyed Susan, are hydroxylated or oxygenated carotenoids called xanthophylls. Brown algae, diatoms, and other algae also possess oxygenated carotenoids.

The pink color of flamingos is due to carotenoids, which the birds obtain directly from the algae they eat. Vitamin A is a product of beta carotene cleavage, or carotenoid cleavage; retinine is an oxidation product of vitamin A. In summary, carotenoids are conspicuous in all photosynthetic organisms, including bacteria, but they are also present, in lower concentrations, in other organisms.

Please discuss the effects of terpenoids on growth in animals.

Most of the studies of terpenoids have considered their effects as feeding deterrents of animals, primarily insects. I would like to point out that this is a very large question, because of the great diversity of terpenoid structures. If we want to remember that steroids are terpenoids, we can mention molting hormones and note the important metabolic effects of steroids.

Are alkaloids present in most flowering plants? How do you explain the recent development and the distribution of these chemicals?

While nearly all known alkaloids are produced by flowering plants, they occur in only about 20 percent of them, primarily dicots.

Alkaloids are a large group of compounds characterized by heterocyclic rings of nitrogen and carbon atoms. They are basic compounds in the sense that they are soluble in acid and take up protons (H^+ ions); hence the name "alkaloid." They are also characterized by being physiologically active. Alkaloids are synthesized via a variety of metabolic routes. Although most are derived from amino acids, some alkaloids are derived from steroids. Alkaloids are not homologous to each other, nor do they have a single origin; rather, this vast group of compounds is defined by their chemical characteristics.

Please discuss the extent of the ability among animals to detoxify phenols, terpenes, and alkaloids.

Animals vary greatly in their ability to detoxify phenols, terpenes, and alkaloids. Briefly, it seems that many animals have, among other mechanisms for dealing with toxicity, mixed-function oxidases (MFOs). The speed with which these enzymes are produced often varies among insects. In generalist-feeding insects, MFOs are induced very rapidly. More specialized feeders, insects that have only one or few food sources, must deal in other ways with toxins from their particular food plants. These mechanisms vary greatly, and depend on the evolutionary history of the animal.

Please discuss the range of alkaloid toxicity in various groups of animals. Are the same alkaloids toxic to some animals and not to others?

Toxicity of alkaloids varies greatly among animals. Nicotine, for example, is used as an insecticide, yet the tobacco horn worm, the cabbage looper, and the tobacco bud worm are all resistant to it. Another group of alkaloids, the pyrrolizidene alkaloids found in ragworts (*Senecio*), are hepatoxins. They are toxic to the liver of mammals, but not to certain moths, and they are used as sex attractants by danaid butterflies.

**How did flowering plants develop terpenes less complex than
those of the earlier vascular plants? Does the biosynthesis of the
longer polymers require more energy? How is it that longer-chain
terpenes are found in the plants and bacteria that appear earlier in
the fossil record?**

In answering why flowering plants developed terpenes simpler than
those of earlier vascular plants, there is the assumption that the
longer polymers would be harder to synthesize and therefore would
take longer to evolve. In fact, it is appropriate to realize that carotenoids and other long-chain terpenoids are absolute requirements for
photosynthesis. Later, certain plants developed the ability to modify
simpler monoterpenes and sesquiterpenes into ecologically useful
compounds. We see this particularly in gymnosperms and angiosperms, which make a great variety of the simpler terpenoids. The
pathway to the production of terpenoids has been a part of metabolism since the Archean eon. The ability to modify terpenoids into
useful, ecologically relevant compounds, as in many gymnosperms
and angiosperms, apparently evolved more recently, during the
Phanerozoic eon.

**What allowed some insects to continue to feed on early plants
even when these plants produced compounds toxic to the insects?**

We can only infer that the same diverse mechanisms that were present in the past are also present today; these include simple behavioral mechanisms such as taste avoidance and more complex
detoxification mechanisms. Some insects evolved specializations in
which they adapted to particular plants and became able to detoxify
the specific compounds present in those plants. As long as the insect
food is limited to those plants, no harm comes to them from these
secondary metabolites. Animal physiological mechanisms for dealing with toxic plants include rapid excretion of toxins, sequestering
of toxins, detoxification mechanisms (including gut microbiota that
can deal with secondary metabolites), surfactants which break down
lipid compounds or solubilize water-insoluble compounds so they
can be excreted, and various degrees of tolerance to toxins.

Insects seem to evolve faster than flowering plants, perhaps because they have more rapid generation times. They apparently have

the ability to produce a great amount of genetic variation from which mechanisms for dealing with plant toxins could evolve.

With what frequency does color vision occur in major plant-eating groups of animals (i.e., insects, birds, and mammals)?

There is color vision in insects and birds. This is obvious from the way flowers attract insect and bird pollinators; it is also suggested by the simple observation that these groups of organisms are often very colorful themselves. Most mammals, except the primates, seem to lack color vision.

Over what span of time did the evolution of semiochemicals (e.g., alkaloids or terpenoids) take place?

How long it takes a metabolic pathway to evolve is a very speculative question. Alkaloids evolved relatively recently within flowering plants—in the Cretaceous period of the Mesozoic era, about 100 million years ago. Some of what are considered the more primitive groups of angiosperms, such as the Magnoliales and Ranunculales, are alkaloid-containing groups, so we assume that certain alkaloids evolved early in the history of the group. Alkaloids are also present in some lycopods and horsetails (but not in ferns), in a few conifers, and in the gnetophytes (e.g., *Ephedra* and *Welwitschia*), yet the vast majority are found in the dicotyledonous flowering plants.

The terpenoid pathway, on the other hand, is extremely old; terpenoids are involved in the electron-transport chain and in photosynthesis. All chlorophyll molecules, for example, contain an isoprene derivative. The ability to make terpenoids must therefore be very old, perhaps as old as life itself.

Please discuss plant-to-plant semiochemical interactions (e.g., chemical communication between alders and willows when attacked by insects). Over what sort of distances can plants communicate via semiochemicals?

The idea of plant-to-plant semiochemical communication (i.e., that trees "talk" to each other) is really an open question now; whether it really occurs is still under consideration. I think the data for it is not terribly strong. Some plants may do it, other plants definitely do not. The best study is that of Rhoades (1983).

Also, keep in mind that no one has looked at interspecific communication, such as willows communicating with alders, as the question asks. Rhoades found that only the near neighbors of the same species of an attacked tree gave any indication of having received any kind of signal. It is an interesting area of research. There is still a fair amount of controversy over whether plants can chemically communicate with each other.

Readings

Fenecal, T. 1982. Natural products chemistry in the marine environment. *Science* 215: 923–937.

Harborne, J. G. 1982. *Introduction to Ecological Biochemistry,* second edition. Academic Press.

Rhoades, D. F. 1983. Responses of alder and willow to attack by tent caterpillars: Evidence for pheromonal sensitivity of willows. In *Plant Resistance to Insects,* ed. P. Heden. American Chemical Society.

Robinson, T. 1979. The evolutionary ecology of alkaloids. In *Herbivores,* ed. G. Rosenthal and D. Janzen. Academic Press.

Swain, T. 1977. Secondary compounds as protective agents. *Annual Review of Plant Physiology* 28: 479–501.

Visser, J. H. 1983. Differential sensory perception of plant compounds by insects. In *Plant Resistance to Insects,* ed. P. Heden. American Chemical Society.

12

Mammalian Evolution: Karyotypic Fission Theory

Neil Todd

Neil Todd has noted correlations between chromosome numbers in living organisms and sudden and impressive episodes of adaptive radiation in mammals in the fossil record. Karyotypic fission is a mechanism of "macromutation"—rapid rather than gradual evolutionary change.

Todd has worked on his concept of karyotypic fission in the context of carnivores, especially dogs and artiodactyls (pigs, camels, horses, and their relatives). His ideas have not been understood or criticized seriously in the mainstream literature of neo-Darwinism. In this chapter he describes the essentials of his theory.

Karyotypic fission is a theory of chromosomal evolution I developed to explain certain phenomena of mammalian evolution. I believe that karyotypic fissioning is a major mechanism of speciation that applies to all tetrapods (amphibians, reptiles, birds, and mammals) and to other species of animals whose sex is determined by a pair of sex chromosomes.

A karyotype is a representation of the mitotic metaphase chromosomes arranged in accordance with a specific convention by homologous pairs. Take, for example, the karyotype of a long-nosed bandicoot, an Australian marsupial (figure 1). The diploid number of chromosomes of the bandicoot is 14. The autosomal chromosomes are depicted in two rows in order of decreasing length, followed by the two sex chromosomes, X and Y. The arms of each chromosome join in a region called the *centromere*. The *kinetochore*, a microtubule-chromatin junction visible by electron microscopy, forms at the centromere region before chromosome segregation during mitosis. Karyotypic fissioning corresponds to the duplication and dissociation of the centromeres

Order: MARSUPIALIA   Family: PERAMELIDAE

Perameles nasuta (Long-nosed bandicoot)

2N=14
male
2N=12M+XY

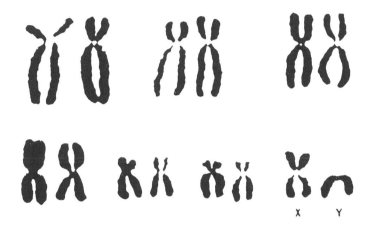

N= haploid number of chromosomes
M= mediocentric chromosome

Figure 1
Karyotype of *Perameles nasuta* (long-nosed bandicoot).

such that both arms of any two-armed chromosome become two
independent chromosomes or genetic linkage groups (figure 2). A
linkage group consists of all the genes or genetic loci on a single
chromosome. Karyotypic fission usually involves the entire chromosomal complement of the cell in which it occurs, although the eventual
karyotype may be modulated by many other factors. Thus, it is not the
karyotypes that actually fission, but rather the chromosomes or linkage
groups from which karyotypes are constructed. Since recognition involves the comparison of karyotypes, the phenomenon is called *karyotypic fission* rather than chromosomal fission. Chromosome fission may

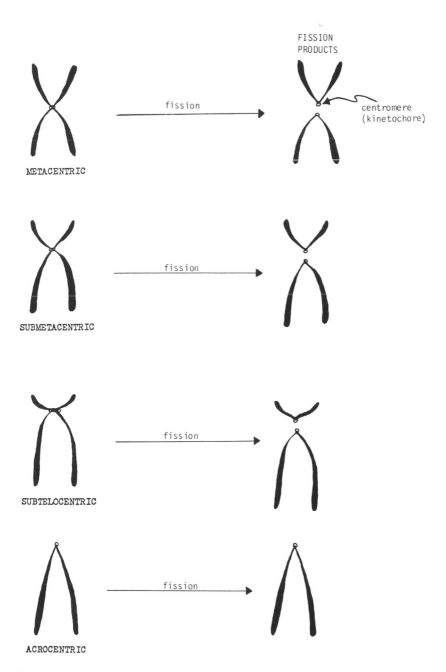

Figure 2
Products of chromosomal fission.

yield various products, as figure 2 shows. If the centromere of a particular chromosome is central, after fission two chromosomes exist where there was originally one. If the centromere is terminal, the chromosome and the linkage relation of the genes on that chromosome remain essentially unchanged.

Hypothetical Phylogenies of Fissioned Karyotypes

Karyotypic fission can be correlated to the phylogeny of mammals as reconstructed from the fossil record. Phylogenies are hypothesized genealogical sequences, often diagrammed as a branching tree, of ancestor-descendant relationships. Mammals first appear at the Triassic-Jurassic boundary, about 180 million years ago. Since then between 20,000 and 40,000 species have evolved, though today there may be only 4000 or more extant species. The reasonably good fossil record and the various kinds of comparative studies allow the overall relationships and pedigrees of many of the surviving lines to be established. Exactly how many species have been karyotyped is not known, but estimates are in the vicinity of 2000 in a good cross-section of mammals. As might be expected, the karyotypes of some groups are much better known than those of others. The diploid number of chromosomes in mammals ranges from 6 to 92. A deer, *Muntiacus muntjak,* is at the low extreme, with a diploid number of 6 in the female and 7 in the male (figure 3); a South American dog, *Atelocynus microtis,* has a diploid number of 74 (figure 4).

Despite this somewhat bewildering range, there are also fundamental similarities among mammalian genomes. These similarities are critical to the understanding and interpretation of the relation of karyotypic fission to mammalian evolution. The total DNA content of the mammalian genome is remarkably constant. DNA content varies from about 85 percent to 105 percent of that of the *Homo sapiens* genome. The sex of any animal is determined by a pair of sex chromosomes. The presence of two X chromosomes determines the sex as female, whereas all males are XY. The X chromosome, although sometimes involved in peculiar translocations, represents approximately 5 percent of the haploid genome in all mammalian species. In most mammals the X chromosome is of medium size and is mediocentric (that is, the centromere is near the middle). From these facts certain inferences can be drawn:

Order: ARTIODACTYLA   Family: CERVIDAE
         (even-toed ungulates)   (deer, antelope)

Muntiacus muntjak (Indian or red muntjac)

2n=♂7, ♀6

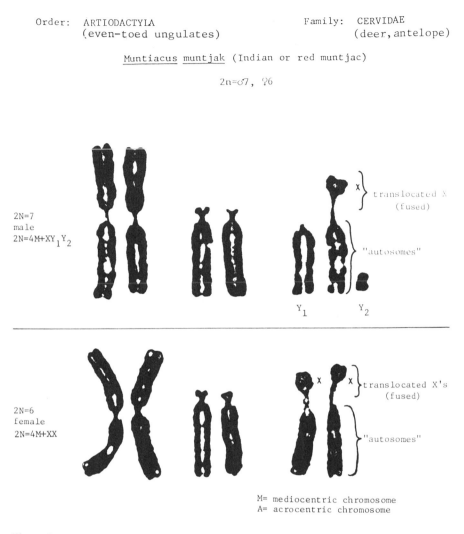

M= mediocentric chromosome
A= acrocentric chromosome

Figure 3
A deer karyotype: *Muntiacus muntjak* (Indian or red muntjac).

First, all differences in mammalian karyotypes result primarily from
the repackaging of a seminal genome, not from the addition or the
subtraction of material. Second, the sex-determining mechanism is
largely refractory to rearrangement—in other words, the X chromosome usually survives and is present in very disparate lineages, no
matter what the diploid number. The Muntjac deer has essentially the
same X chromosome (although it is attached to an autosome) as the

Order: CARNIVORA   Family: CANIDAE
                                                                (dogs)

Atelocynus microtis (round-eared dog or small-eared dog)

2N=74
female
2N=72A+XX

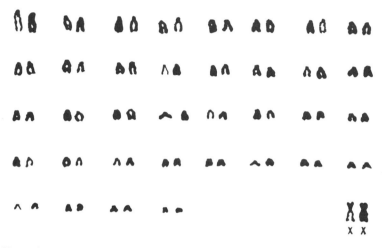

M= mediocentric chromosome
A= acrocentric chromosome

Figure 4
A dog karyotype: _Atelocynus microtis_ (round-eared or small-eared dog).

South American dog; the total amount of DNA in these two karyotypes is not grossly different.

In the construction of a phylogeny of a particular group, it is possible to proceed to an analysis of the distribution of the karyotypes in all the extant derivatives within this group. In a hypothetical phylogenetic tree, a common ancestor is the starting point from which lineages (including extinct ones) can be traced (figure 5). The aim is to reveal the sequence of events (especially adaptive radiations or cladistic episodes) in each lineage—in short, the emergence of diverse species from an ancestral form. One descendant (the one on the left in figure 5) has

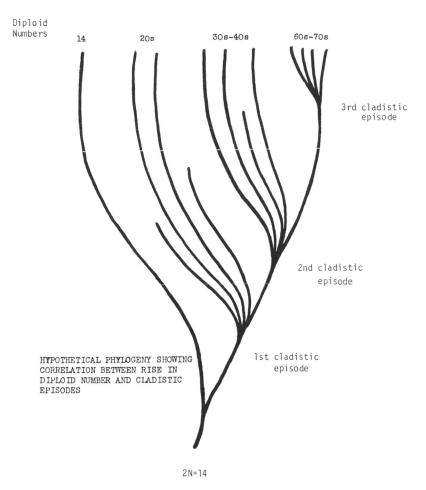

Diploid
Numbers

14   20s   30s-40s   60s-70s

3rd cladistic
episode

2nd cladistic
episode

HYPOTHETICAL PHYLOGENY SHOWING
CORRELATION BETWEEN RISE IN
DIPLOID NUMBER AND CLADISTIC
EPISODES

1st cladistic
episode

2N=14

Figure 5
Hypothetical phylogeny (based on chromosome numbers), showing correlation between
diploid values and cladistic episodes.

more or less extended into the present, and is not a derivative of any
large adaptive radiation. Although this species may be quite different
from the ancestral species in terms of particular specializations, it has
had no tendency to diversify. An episode of adaptive radiation appearing in this line is termed the *first cladistic episode* or the *primary radiation;*
a second cladistic episode, if recognizable, is called the *secondary radiation,* and so on. If the diploid number of the species descending directly
to the present is, for example, 14 ($2n = 14$), then we assume this is the

ancestral diploid number. As we look at derivatives of a single cladistic episode, we find the diploid number elevated to the mid 20s. If a derivative line descends through two adaptive radiations, the diploid number will be higher still, roughly in the high 30s or the 40s. If a third episode of adaptive radiation in this line of descent emerges, diploid numbers in the 60s or the 70s, or higher, may be found. Today such patterns of diploid numbers are known to exist in numerous lineages of mammals, although never as ideally as in this hypothetical construction.

The ploidy of $2n = 14$ postulated in the last example was not a random choice. Marsupials are found in Australia and South America, and the few North American representatives are descendants of South American Pleiocene migrants. The extant South American marsupials are considered vestiges of the once-diverse marsupial fauna, which has been reduced to a dozen or so species by competition with placentals. Australia and South America were once joined with Antarctica in a large continental mass known as Gondwanaland. Since the breakup of Gondwanaland, South America and Australia have been drifting apart for at least 100 million years. The long-term separation of the continents explains the major features of the present distribution of the marsupials, including their karyotypes. Even though the marsupials in South America and Australia are morphologically very different, they have virtually identical karyotypes, with a diploid number of 14. Diploid numbers up to 30 do occur, but these karyotypes do not much resemble one another either within or between the two continental faunas. The similarity of ploidy is convincing evidence that the primitive diploid number for marsupials was 14. In placentals a diploid number of 14 has been reported for a peculiar little group: the Macroscelididae (better known as elephant shrews, though they are neither elephants nor shrews). This report needs confirmation but would be a very important finding for phylogenetic placement, for the Macroscelididae lie somewhere between marsupials and placentals, albeit much nearer the latter. A diploid number of 14 occurs in the South American rodent *Akodon*, although several other species in the same genus have diploid numbers in the 50s. Finally, in a modified form, the karyotype of *Muntiacus muntjak* (figure 3) can be "dissected" to give the 14 elements found in the basic marsupial karyotype. Hence, though $2n = 14$ is not common in placental mammals, it is found in diverse representatives. That the earliest marsupials had a diploid chromosome number of 14 is certainly an appropriate hypothesis, and the similar diploid number of diverse placental mammals indicates to me that 14 is fundamental to all mammals as the ancestral mammalian karyotype.

Chromosome Polymorphisms May Lead to Speciation

A more detailed examination of chromosome terminology may clarify the issues. Chromosomes are designated and distinguished according to the ratio of their arm length, and are placed in four classes (figure 6). If the centromere lies at the center, the chromosome is called *metacentric*. If the centromere lies nearer the center than the end, the chromosome is *submetacentric*. When the centromere is nearer the end than the middle, the chromosome is *subtelocentric*. If the centromere is terminal, the chromosome is *acrocentric*. Acrocentric (derived from the Greek *acros*, meaning "edge") is preferred to telocentric (derived from *telos*, meaning "end"), as it is difficult in practice to determine that there are absolutely no functional genetic loci beyond the centromere.

Karyotypic-fission theory is concerned only with two broad groups of chromosomes: mediocentric (the three types with nonterminal centromeres) and acrocentric. After fission, a mediocentric chromosome yields two independent linkage groups; an acrocentric chromosome is still acrocentric and does not create new linkage groups. In other words, fission only elevates the diploid number in proportion to the mediocentric chromosomes present in the karyotype. In any population one may find polymorphisms for these various classes of chromosomes. One individual may have a large mediocentric chromosomal pair, whereas another may have two smaller matching acrocentric pairs. This in itself does not provide evidence of chromosome fission or fusion, although a careful analysis of banding patterns might reveal differences. Breeding these animals, however, provides a definitive test. Domestic pigs generally have a diploid number of 38, but some have 37 or 36. Karyotype comparison shows the presence of a mediocentric-acrocentric linkage polymorphism group (figure 7). Pigs with a diploid number of 36 have a pair of medium-size mediocentric chromosomes. Those with 37 have one such chromosome and two smaller acrocentrics that represent the same linkage group. Those with 38 do not have the medium-size mediocentric linkage at all but are called *homomorphic* because they have two pairs of small acrocentrics.

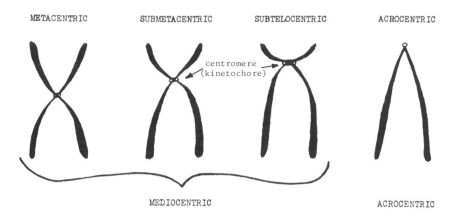

Figure 6
Chromosome nomenclature.

Remarkably, the differences are transmitted in a simple Mendelian fashion, with no evidence of reduced viability or fertility between any of the morphs. Two pigs with diploid numbers of 37 (heteromorphic) can interbreed and will produce offspring with 36, 37, and 38 chromosomes in a 1:2:1 ratio. The inescapable conclusion is that the two smaller acrocentrics function as one linkage group in the heteromorphic individual.

A second kind of polymorphism commonly found in some groups involves a transformation of a mediocentric chromosome to an acrocentric one (figure 8). This polymorphism is a particular linkage group in the form of either an acrocentric or a mediocentric chromosome. The transformation is caused by *pericentric inversion;* that is, the centromere changes position. Pericentric inversions are widespread—for example, in the deer mouse *Peromyscus* 20 or more chromosomal pairs are involved. Evidence suggests no physiological impairment or infertility when mating individuals have such different karyotype morphologies. Indeed, many of these morphs are distributed in clines (that is, in gradients from high to low frequencies along a geographic transect through the population). Pericentric inversions do not change the diploid number. Because of fission, populations of *Peromyscus* also have differences in diploid number. Neither fission polymorphisms nor pericentric inversions appear to create any problems on their own; however, the simultaneous occurrence of a fission event and a pericentric inversion seems to severely limit interbreeding. One can imagine

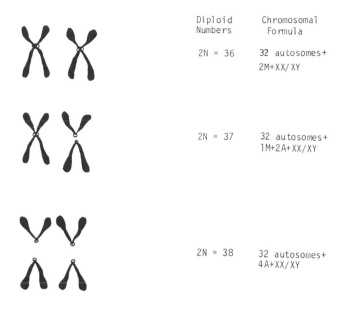

	Diploid Numbers	Chromosomal Formula
	2N = 36	32 autosomes+ 2M+XX/XY
	2N = 37	32 autosomes+ 1M+2A+XX/XY
	2N = 38	32 autosomes+ 4A+XX/XY

M= mediocentric chromosome
A= acrocentric chromosome

Figure 7
Chromosomal polymorphisms: fission polymorphs in pigs.

the consequences of breeding individuals with two different polymorphisms (figure 9). On the left side of the illustration is an individual with the polymorphism for the pericentric inversion (as seen, for instance, in *Peromyscus*). On the right side is an individual with a polymorphism for the fission product (as seen in the case of the domestic pig). Individuals with either of these two types have no overt physiological problems; however, if they mate with each other, among the offspring types will be the one portrayed at the bottom. This individual will tend to be infertile because it will have great difficulty producing gametes; the two centromeres of the fission product will not align properly during meiotic synapsis. Individuals with this karyotype will produce a high percentage of inviable gametes. Thus, this karyotype will tend to be lost in the population.

Central to our inquiry are the questions of where karyotypic fission occurs and how it becomes established in a population. There are various scenarios. One involves an individual who carries in his or her

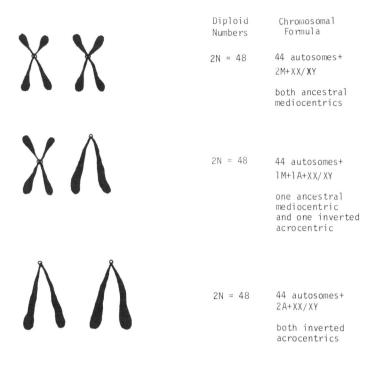

	Diploid Numbers	Chromosomal Formula
	2N = 48	44 autosomes+ 2M+XX/XY both ancestral mediocentrics
	2N = 48	44 autosomes+ 1M+1A+XX/XY one ancestral mediocentric and one inverted acrocentric
	2N = 48	44 autosomes+ 2A+XX/XY both inverted acrocentrics

M= mediocentric chromosome
A= acrocentric chromosome

Figure 8
Chromosomal polymorphisms: pericentric-inversion polymorphs in the genus *Peromyscus* (deer mouse).

germ line a fully fissioned karyotype. This individual might introduce a fissioned complement of chromosomes into the population again and again. Such a case has been documented in a population of lizards. In my view there tends to be a selective advantage in breeding between animals with the same karyotypes—although breeding between morphs with chromosome alterations occurs, it may lead to reproductive impediments. Many mammals exhibit some kind of mating preference. Mole rats (*Spalax*) seem to somehow differentiate between potential mates with 56 or 58 or 60 chromosomes. Having this faculty for distinction, they tend to mate with their own kind. With a mechanism for generating new karyotypes and the ability to somehow distinguish karyotypes, single (i.e., interbreeding) populations can

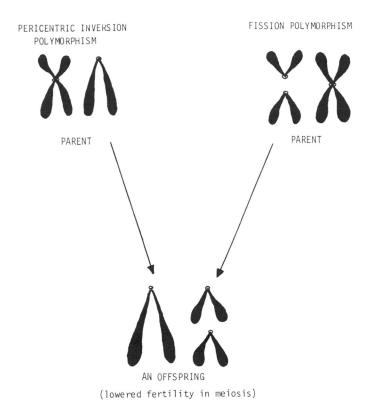

PERICENTRIC INVERSION
POLYMORPHISM

FISSION POLYMORPHISM

PARENT

PARENT

AN OFFSPRING
(lowered fertility in meiosis)

Figure 9
Chromosomal polymorphisms: result of mating individuals with fission polymorphism and pericentric-inversion polymorphism.

suddenly be fractured by these chromosomal changes. In a population in which a pericentric inversion polymorphism is distributed as a cline, fission may occur in one portion of the population in which the inversion is absent and spread until individuals encounter others with the pericentric inversion. At this point interbreeding breaks down and the animals, if they have the faculty for making a distinction, rapidly avoid mating with individuals with a karyotype distinct from their own. In this manner two distinct, reproductively isolated populations—in other words, two diverse species—arise. Other selective pressures, such as pre-mating isolation or mating behavior, can reinforce this isolation.

Evolution of Dog Genera: Application of Karyotypic-Fission Theory

The karyotypes of modern dogs provide a useful example of the explanatory power of karyotypic fission in the resolution of phylogenies. Table 1 details the karyotypes of the extant genera of dogs: *Canis* (the domestic dog, the wolf, the coyote, and the jackal), *Chrysocyon* (the maned wolf), *Atelocynus* (the small-eared dog), *Dusicyon* (several closely related subspecies, including the Andean wild dog known as the "lobo"), and *Urocyon* (the grey fox). Today *Canis* is distributed over much of the world; *Chrysocyon, Atelocynus,* and *Dusicyon* are confined to South America, and *Urocyon* to North America. The fossil record indicates that North America was the ancestral home of the stocks that gave rise to all these modern genera. A hypothetical array of karyotypes, all having a diploid number of 42 but differing in the proportions of mediocentric and acrocentric chromosomes, can be constructed (table 2). Fission of one of these putative karyotypes could have led to the present karyotype of *Urocyon* as follows:

$$2M + 24M + 14A + XX/XY \xrightarrow[\substack{\text{fission +} \\ \text{retention of} \\ \text{two ancestral} \\ \text{mediocentrics}}]{} 2M + 48A + 14A + XX/XY.$$

In table 1, the high number of chromosomes and the complete absence of mediocentric chromosomes (except for one pair in *Urocyon*) suggest a recent fission event. My assumption is that acrocentric chromosomes are gradually converted to mediocentrics, since a mediocentric linkage is presumably more stable during mitosis than an acrocentric. This stability is a function of the length of the chromosomes; the location of the centromere will have little to do with the likelihood of breakage along a small chromosome during mitosis. Here the aim is to deduce the karyotype and the diploid number of the last common ancestor of this derivative group. My estimate is that the ancestral diploid number was not lower than 40; this is based on the fact that it requires 38 mediocentrics plus the XX/XY sex chromosomes to yield 76 acrocentrics plus the XX/XY chromosomes in a single fission event. However, for reasons that will become clear, I am going to designate a diploid number of 42 as the ancestral diploid number.

Table 1
Extant genera of dogs and their karyotypes.

Canis	2N = 78 = 76A + XX/XY
Chrysocyon	2N = 76 = 74A + XX/XY
Atelocynus	2N = 74 = 72A + XX/XY
Dusicyon	2N = 74 = 72A + XX/XY
Urocyon	2N = 66 = 2M + 62A + XX/XY

All of what I believe were the four karyotypes of the five incipient genera have the same diploid number (42), but there are differences in the morphology of the chromosomes—in other words, in the karyotypes (table 2). The group is an interbreeding population with a polymorphism for several mediocentric-acrocentric pairs, not unlike polymorphisms reported for several extant species. If the migration of the populations coincided with this polymorphism, then we have constructed a cline found in extant populations of dogs. A fission event occurring in the incipient *Canis* population, if introduced into each of the other incipient groups, yields exactly the karyotypes of the five modern genera. The only wrinkle in this example is in *Urocyon*, where in addition to the fission a single pair of ancestral mediocentric chromosomes must have been conserved.

I hypothesize that the steps that led to the *Urocyon* karyotype involved the nonfission of two mediocentric chromosomes of the original 26; these chromosomes remained in the ancestral state, as was diagrammed above. This hypothesis could be dismissed as simply the juggling of chromosome numbers to achieve the desired outcome were it not for the existence of *Nyctereutes vivarinus,* the raccoon dog. The karyotype of *N. vivarinus* includes 40 mediocentrics and the two sex chromosomes. Of these mediocentrics, one assumes that 14 of the original autosomes were acrocentric but have subsequently been converted by pericentric inversion to mediocentrics. Evidence from the fossil record shows that the ancestral stock of modern dogs crossed the Bering Strait into Asia. The Asian population of dogs subsequently became geographically isolated when the Bering land bridge was submerged. Thus, *Nyctereutes*, according to the facts of zoogeography, is the ancestral form. I admit that this example of the evolution of dog genera is exceptional in terms of the completeness of the fossil record, the knowledge of migration, and the survival of an ancestral

Table 2
Hypothetical ancestral karyotypes.

2N = 42 = 36M + 4A + XX/XY	Incipient *Canis*
2N = 42 = 34M + 6A + XX/XY	Incipient *Chrysocyon*
2N = 42 = 32M + 8A + XX/XY	Incipient *Atelocynus, Dusicyon*
2N = 42 = 26M + 14A + XX/XY	Incipient *Urocyon*

representative; nonetheless, it establishes confidence in the relevance of paleokaryology.

Karyotypic fission also offers an explanation for the observation that evolution, contrary to expectations based on the theory of population genetics, appears to occur in spurts. In other words, morphological change and diversification occur suddenly, not gradually, and are followed by a period of relative stasis (Gould and Eldredge 1977). The proposition of punctuated evolution was long held to be an artifact of incomplete sampling; however, the theory of punctuated equilibria has been sufficiently documented to at least call into question the uncritical acceptance of gradualism. Recent findings and interpretations of the hominid fossil record indicate that spurts are real. No adequate or satisfying genetic explanation for the sudden emergence of new traits has been offered, but I submit that—at least for mammals—karyotypic fission is a good candidate, for it explains two fundamental aspects of evolutionary events. First, by providing a plausible mechanism for the generation of reproductively isolated populations, karyotypic fission explains the rapidity of speciation. Second, any event of karyotypic fission imposes a period in which derivatives have no potential for repeating the process. In other words, fission dictates a pause. During this pause, surviving lineages must accumulate mediocentric linkages before they can enter another phase of fission. The time required for such repotentiation can reasonably be expected to be millions of generations.

Karyotypic fission addresses some rather controversial evolutionary topics. It provides a mechanism for sympatric speciation. It addresses the topic of punctuated equilibria in evolution. It provides a potential explanation of adaptive radiations. In general, karyotypic fission seems a very promising approach to the analysis of tetrapod evolution. To date, karyotypic-fission theory has been applied only in limited cases: Canidae (Todd 1970), primates (Stanyon 1983; Giusto and Margulis

1983), artiodactyls (Todd 1975), and, to a limited extent, perissadactyls and their relatives (Todd 1975). A brief overview of other groups suggests that they are amenable to this experimental approach. I am sufficiently confident in the power of karyotypic-fission theory to explain the interrelations of extant species that I believe it is possible to reconstruct the evolutionary history of these species even in the absence of a fossil record. This may be particularly useful for those groups for which fossil records are poor, such as bats and whales. Karyotypic fission theory has been applied to a limited number of animals by a limited number of workers. Fewer than 100 of the 4000 species of mammals have been analyzed. Ample opportunity exists for work relating to the remaining thousands of species. Numerous projects await our attention.

Readings

Gould, S. J., and N. Eldredge. 1977. Punctuated equilibria: The tempo and mode of evolution reconsidered. *Paleobiology* 3: 115–151.

Giusto, J. P., and L. Margulis. 1981. Karyotypic fission theory and the evolution of old world monkeys and apes. *Biosystems* 13: 267–302.

Giusto, J. P., and L. Margulis. 1983. Karyotypic fissioning (letter to the editor). *Biosystems* 16: 169–172.

Kolnicki, R. Karyotypic theory applied: Kinetochore reproduction and Lemur evolution. *Symbiosis* 26: 1–19.

Stanyon, R. 1983. A test of the karyotypic fissioning theory of primate evolution. *Biosystems* 16: 57–63.

Todd, N. B. 1970. Karyotypic fissioning and canid phylogeny. *Journal of Theoretical Biology* 26: 445–480.

Todd, N. B. 1975. Chromosomal mechanisms in the evolution of artiodactyls. *Paleobiology* 1: 175–188.

13

Environmental Pollution and the Emergence of New Diseases

Jonathan King

Inhibition of a population's growth via behaviors by that same population that produce waste and toxins has many precedents in the evolution of microbial, plant, and animal life. The present situation is discussed here by Jonathan King, a member of the board of directors of the Council for Responsible Genetics (Cambridge, Massachusetts) and a professor of molecular biology at the Massachusetts Institute of Technology.

Organisms encounter a wide variety of chemicals, minerals, and macromolecules during their life cycles. Until about 10,000 years ago, the great majority of these substances were products, by-products, and breakdown products of other organisms or the results of weathering and other physical processes. Most of the evolution of primate and human physiology occurred under these conditions. With the evolution of speech, cooperation, and culture, human societies moved to a higher level of organization and complexity, giving birth to the agricultural and industrial revolutions.

Environmental Contaminants

Mining, metallurgy, manufacturing, and intensive industrialization have resulted in a vast increase in the variety of chemicals and minerals to which nonhuman organisms and human beings are exposed. In the United States, this exposure increased dramatically with the growth of the petrochemical and agrochemical industries after World War I. One consequence has been the emergence of new diseases associated with physiological inability to cope with the toxic affects of exposure to some of these substances. Poisoning from environmental pollution occurs in micro-organisms, in plants, and in animals. Often it is ignored or undetected. In 1962 Rachel Carson's book *Silent Spring* called attention to the loss of bird life caused by the use of insecticides. Since then, the public's attention has been drawn to birth defects in frogs, to tumors in flounder and other game fish, to coral reef diseases in the Florida keys, and to the massive destruction of marine life at sites of oil spills.

Ever since the cholera epidemics of the late nineteenth century we have been sensitive to pollution of drinking water by bacteria and viruses. Pathogenic strains of *E. coli, Salmonella typhimurium* (which causes food poisoning), *Giardia,* and polio virus are just a few of the infectious agents that sanitary systems attempt to exclude from drinking water. In this chapter I will focus on agents which are not infectious, and which in fact may be regarded as inert, but which nonetheless can cause serious disorders in humans, animals, and plants.

Some of the chemical and mineral substances of concern are present naturally in the Earth's crust but are normally locked away from human exposure. Examples include mercury, lead, cadmium, and uranium. Only with the development of mining and metallurgy were these substances isolated from their mineral context and brought into the human ecosystem.

Another set of substances were not present in significant quantities anywhere on Earth until industrialization. Examples include many synthetic organic and inorganic chemicals (chlorofluorocarbons, polybrominated biphenyls, tetraethyl lead), many of their polymerized forms (nylon, plastics, zeolites), and plutonium. The American Chemical Society lists more than 4 million distinctive chemical compounds, most of which have been synthesized only in the last 100 years. From 60,000 to 70,000 of these 4 million compounds are believed to be in regular use in industrial societies. Most of these compounds begin their existence in the human environment in very concentrated form at some very distinctive industrial site (called a *point source*). They then are distributed or dispersed through diverse pathways. Some are in our foodstuffs, some in the ceramics of our coffee mugs, and some in hair dyes, in gasoline, in paints, and in chemicals used in laboratories. We are aware of some of the more unpleasant dispersed forms, such as smokestack emissions and automobile exhausts.

A third class of substances may have been present at very low levels at one time, but their levels have been increasing sharply since industrialization began. (For example, a number of naturally occurring phenolic compounds have been found in far higher local concentrations where plastics are produced and where certain other kinds of manufacturing take place.) Many of these substances are brought into existence at local and distinctive sites. Plutonium is created primarily at nuclear reactors used in the manufacture of nuclear weapons. Polychlorinated biphenyls are brought into existence at a limited number of manufacturing plants producing these chemicals for the electrical equipment industry. Mercury as a purified metal is produced at a limited number of smelters built for that purpose. Such substances end up distributed very broadly and irregularly through the environment—for example, lead from paint pigments and gasoline exhausts. Nonetheless, though these substances may end up dispersed through the overall environment in very low concentrations, there are point sources where they occur in concentrated form, and where humans, animals, and plants may have been exposed to very high concentrations. We will return below to consider some consequences of this.

At present the dispersion of these agents from their point source are rarely systematically tracked. Despite the enormous body of data on the toxicity of mercury, the distribution of mercury from the smelters where the metal is initially produced is not followed. Corporations of course have production and shipping records, but these are not generally publicly available. Even in cases where there has been major legal action, such as the manufacture of asbestos products by the Johns Manville Corporation, the final distribution of the materials is often not publicly known. In an important exception, much better records are kept of radioactive compounds produced under the federal government's supervision, though they may not be available to a local community.

The Etiology of Disease

Physicians use the term *etiology* for the cause or origin of a disease. We need to be able to identify etiologic agents which are of human origin in order to stop introducing them into the environment and to remove those already released. This is a crucial step in the long-term prevention of human disease and alleviation of human suffering. It is also important to identify introduced agents that harm animal, plant, and microbial species, in order to sustain the environment that supports us.

The toxicity of certain classes of contaminants has been known since ancient times. The Romans mined lead and used it in paint for their ships. Pliny noted that the workers put masks over their faces to prevent being poisoned by inhaling the powders. Long before there were scientific and medical journals devoted to these subjects, people living close to point sources of contamination knew of them.

How do we determine the effects of more recent environmental additives on animals and plants? For most animals, plants, and microorganisms there are few systematic efforts to study these processes. However, if the organisms affected are economically important, such as major crops or dairy herds, the economic loss of disease may result in efforts to track down the cause. For humans, epidemiologists study the relation between forms of ill health and identifiable environmental variables. Unfortunately, the distributions of the great majority of organic chemicals are not known. This makes it difficult to track down these relationships by the available experimental means. In addition, today humans are exposed to thousands of xenobiotics (compounds of documented external origin), and it is difficult to disentangle the effects of multiple exposure.

However, the origins of many of the materials brought into existence at mines, smelters, factories, and petrochemical refineries are identifiable. They may not be known to the general community, but they are often known to those who work there and live nearby. The exposure of local humans, plants, and animals to toxic substances at their point of production is often unambiguous.

Animals and plants can be important indicators of trouble. For example, tumors in flounder caught in New Bedford (Massachusetts) Harbor, whose waters are contaminated by polychlorinated biphenyls and other aromatic hydrocarbons from a local electrical plant, were noticed by local fisherman. Such examples of results of exposure at relatively concentrated point sources provide a model for the causative relationships between exposure to pollutants and disease. Major exceptions are mice, rats, and rabbits, which are used in toxicity studies in the pharmaceutical industry and in related industries. There is a very large body of data—some published in the scientific literature, others sitting in notebooks in pharmaceutical industry laboratories—identifying the toxic effects of various compounds on lab rodents.

The first legislation that attempted to prevent unnecessary exposure of people to poisons was the Pure Food and Drug Act (1906). A variety of legislative acts passed in recent decades require testing of certain classes of substances for toxicity and attempt to limit their introduction into the environment. Two such acts are the Toxic Substances Control Act (TOSCA) and the Occupational Safety and Health Act (OSHA). The Clean Water Act and the Clean Air Act set limits on release for substances whose toxicity has been established. These legislative initiatives resulted from social and political efforts by constituencies trying to protect environmental and human health.

Case Studies of Diseases Due to Environmental Contamination

Rickets among Child Laborers

Among the first human diseases whose etiologies were clearly identified were metabolic disorders due to famine or to nutritional deficiencies. Scurvy, common among sailors and seamen deprived of fresh fruit, is due to the absence of vitamin C. Vitamin C is required for collagen metabolism, and its absence results in failures of wound healing. Beriberi and pellagra are vitamin B deficiencies associated with inadequate diets. In the nineteenth century a condition of skeletal deformities and bone brittleness called rickets was common among children in Great Britain. It is caused by deficiency in vitamin D.

We humans can synthesize our own vitamin D in the presence of sunlight. People who work outdoors, as nearly all humans and their ancestors did from 5 million to about 500 years ago, never lacked vitamin D. This changed with the transition from hunting and gathering and farming to indoor work in factories and other buildings. An unfortunate aspect of the industrial revolution and the wage system was the intensification of child labor. In textile mills in Britain, children worked 12–14 hours a day deprived of sunlight and lived in crowded cities where coal was burned and the atmosphere was obscured by soot (coal dust and other products of combustion). During this period, rickets appeared in the children of families who had not experienced such disease before they left the farm for the factory.

Vitamin deficiencies respond dramatically to correct diet, so we don't think of rickets and the others as an environmental disease. Yet, since vitamin D synthesis depends on long-wave ultraviolet light passing through the skin, the occlusion or absence of sunlight in a child's environment causing this deficiency qualifies rickets as an environmental disease. Rickets and other vitamin-deficiency diseases are seldom encountered in Western societies these days.

Heavy Metal Poisoning

The development of civilization is intimately associated with extraction and smelting of metals, including copper, iron, mercury, and chromium. These naturally occurring metals are locked up in the rocks and minerals of the Earth's surface, where humans are not exposed to them in high concentrations. This relationship is evident in figure 1, which plots the mining and extraction of copper from about 5000 years ago up to the present. The peak during Roman times is correlated with copper's use in coins. A decline follows, after which another peak appears during the development of Chinese culture. The beginning of modern times shows a very sharp increase. Today copper is produced in the millions of tons per year and is used in all manner of electrical devices. As figure 2 shows, production of copper, lead, and zinc has increased rapidly since the late nineteenth century. Millions of tons are introduced into the human environment per year.

The history of mercury poisoning is particularly instructive. A major material in the nineteenth-century hat industry was felt. The felting process requires the use of high concentrations of the potent neurotoxin mercury, and exposed hat workers suffered neurological damage. Their hands and feet shook, and muscle spasms in their mouth and tongue led to difficulties with pronunciation. This affliction was known as "the Danbury Shakes" after Danbury, Connecticut, when that town was the center of the hat industry.

Minamata Disease, not originally diagnosed as heavy metal poisoning, was first described in 1953, when an outbreak occurred near Minamata Bay in Japan. The Chisso Company, a manufacturer of polyvinylchlorides, discharged mercury-contaminated solutions into a lake. The fish of the lake accumulated and concentrated mercuric compounds in their tissues. Suffering especially from mercury poison-

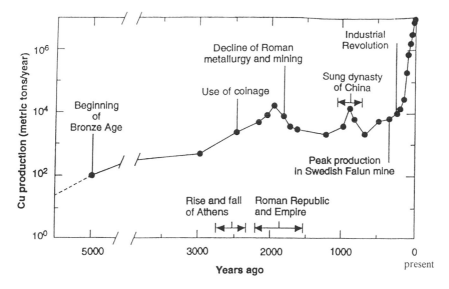

Figure 1
Rate of copper production during the past 5000 years (reconstructed from review of literature).

ing were the poorer people around the lake whose diet was mostly local fish.

In the United States, the major form of metals poisoning has probably been lead poisoning. Lead, besides its effects on neurological processes, causes anemia in some people by interfering with enzymes in heme biosynthesis. (Heme, a ring compound containing iron, nitrogen, and carbon, is a part of the hemoglobin molecule that binds oxygen.) Added to almost all the gasoline sold in the United States for more than 30 years to keep engines from "knocking," lead was steadily introduced into the environment through fuel combustion. It accumulated on roads, on curbs, and in soil. Since 1970, when restrictions on the sale of leaded gasoline were initiated in the United States, the levels of lead on pavement and in the atmosphere have declined.

Between 1920 and 1940, white and light-colored house paints were routinely made with lead pigments. Over time, paint dries out and flakes off. Children, especially in areas with poorly maintained housing stocks, ate or chewed on the attractive, slightly sweet-tasting paint chips, and lead poisoning ensued. Low-level lead poisoning, especially

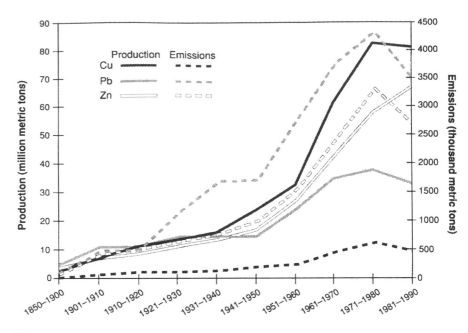

Figure 2
Recent historical changes in mine production and anthropogenic emissions of trace metals to atmosphere (based on Nriagu 1996).

in the young, results in derangement and neurological damage. Lead paint that has not yet been removed remains a problem today.

Suburban houses all over the United States have outdoor decks constructed of pressure-treated wood, which often is impregnated with arsenic compounds. We know little about the release of arsenic from treated wood as the years pass, but certainly arsenic is being released (in some cases "innocently" by carpenters and homeowners as they cut or burn the wood).

Synthetic Dyes and Bladder Cancer

In the United States and in Western Europe, the major inducers of premature death among adults are heart attacks, strokes and other circulatory problems, and cancer. The etiologies of these common ailments are not as well known as those of metal poisoning and bacterial infections. Often they are presented as mysterious and obscure, or are attributed to genetic makeup, diet, or bad luck. Many such factors certainly operate. But if we have enough data to examine the environmental point sources of relevant chemical compounds, in places where people are exposed to environmental contamination at much higher rates of concentrations than is the general population, we can find many clues.

The biology of cancer invariably involves the transformation of normal cells to cancer cells. The cells continue to divide when they should not, forming a tumor or spreading to other parts of the body. A tumor is composed of cells that are the progeny of division of the earlier cells, so that at the cellular level the property is inherited from the parent cells. In general, cancer cells have sustained damage to their genes and chromosomes, so their cell division is poorly regulated by their internal and external environments. Genes that when damaged lead to cell transformation are referred to as *oncogenes*. The term is slightly misleading: cancer is not caused by these genes; it is caused by *damage* to the genes, and by the agents responsible for the damage.

Cancer, in general, is not passed on from individual to individual. One does not "catch" cancer. Cancer does not behave epidemiologically like an infectious disease. Although tendencies toward certain types of cancer may appear more frequently in certain families, most cases of cancer are not inherited in a predictable genetic fashion. However, if the germ cells are damaged, this will be passed down to children, and families in which this has occurred are important subjects for scientific investigation.

Much experimental evidence shows that one etiologic agent in the transformation of a healthy cell to a cancer cell is the accumulation of mutations in genes controlling cell division. Chemicals that cause alterations in DNA are called *mutagens*. If the damage leads to cell transformation and cancer, they are referred to as *carcinogens*.

The first dyes were derived from plants. Since the early years of the nineteenth century, however, reactive coloring compounds have been produced by the chemical industry. They are still in widespread use. First called "coal-tar dyes," many of them turn out to be aromatic amines.

Around the end of the nineteenth century, workers employed in the synthesis and production of aniline dyes in the Rhine Valley of Germany showed a rather high incidence of an otherwise rare bladder and urinary-tract cancer. After World War I the dye industry expanded to Great Britain, where again employees in the industry seemed especially afflicted with bladder tumors. Later, this industry spread to the United States.

Aniline dyes have a distinctive chemical structure. Figure 3 shows a set of structures of the aromatic amines. One is benzidine. Many others, mostly containing aromatic rings and amino groups, involve two, three, or four carbon-rich rings hooked together. Wilhelm Hueper of the National Cancer Institute showed experimentally in the 1940s that dogs exposed to such aromatic amines developed bladder cancer. Epidemiological data indicate that they also cause bladder cancer in humans.

The incidence of bladder cancer by county in the United States provides the crucial clue. New Jersey is a "hot spot," and Salem County has the highest concentration of bladder cancer in the United States. There is nothing especially distinctive about the people who live in Salem County, New Jersey. All kinds reside there, including recent immigrants. But when these data were collected, 25 percent of the people in Salem County were employed in the chemical industry. Some were employed directly in the manufacture of aromatic dye compounds. A retrospective study revealed that 21 percent of the people who had worked in one New Jersey dye factory suffered from bladder tumors.

Workers "upstream"—in a factory that produces benzidine and other such chemicals—are clearly exposed to much higher concentrations than individuals who use the chemicals "downstream" (as, for example, in hospitals, where benzidine is used in low concentrations in the testing of urine for the presence of hemoglobin).

The New Jersey factory referred to above operated before the passage of the TOSCA, OSHA, and Clean Water acts. Any residual amines in its waste effluent probably moved into soil and groundwater. They were also at a much higher concentration in the wastewater from the plant than in the groundwater. But what happens to solutions made up using benzidine after a hospital technician performs the hemoglobin urine test? Some may be flushed down sink drains, some may be incinerated in trash. What happens to benzidine left in a bottle after a hospital closes? Benzidine used in hair dyes probably issues from beauty parlors across the country in wastewater.

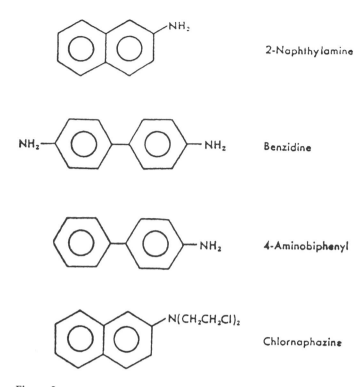

2-Naphthylamine

Benzidine

4-Aminobiphenyl

Chlornaphazine

Figure 3
Aromatic amines.

Aromatic amines are also used as anti-oxidants (to prevent certain unwanted chemical reactions with oxygen in the air) in the rubber industry. The compounds used in the midwest and the south, and probably in other parts of the United States, have probably influenced the incidence of bladder cancer. Exposure to much lower concentrations of these compounds (which we know to be potent carcinogens at high levels) is hard to assess. Although it is easy to make a clear case about bladder cancer in individuals exposed to high concentrations, the lower frequencies in counties where the concentrations are lower may, of course, be due to exposure to aromatic amines dispersed through the ecosystem. Though the precise ecological effects of aromatic amines in the environment are difficult to assess and to define, the existing data suggest the utmost caution.

Combustion generates a diverse class of products, one group of which are polyaromatic hydrocarbons. We are all familiar with cigarette smoking—the inhalation of a complex mixture of combustion products containing well-defined carcinogens that increase the probability of tumors of the epidermal tissues lining the lungs. Tobacco smoke usually isn't thought of as an environmental carcinogen, but of course it is, due to industrial activity of the tobacco industry.

Figure 4 shows the structures of three aromatic compounds present in the combustion of gasoline, coal, or other petrochemicals. In the air they are assessed as industrial pollutants. The relationship between petrochemicals and cancer was first established in the late nineteenth century when Sir Percival Potts, a physician, noted an unusually high incidence of scrotal tumors in young men who had been exposed to tars while working as chimneysweeps in London.

Workers in power plants, in coke ovens, in steel factories, and in poorly ventilated garages are, of course, exposed to higher levels of polyaromatic hydrocarbons than the population at large. In the few studies of coke-oven workers that have been undertaken, these workers showed higher incidences of a variety of cancers. In tests with laboratory rats and other mammals, the aromatic amines behave as skin carcinogens. Some cause mammary-gland cancers in lab animals. Because of their ubiquity and their indirect action, the ring organic compounds can't be easily tracked. Nonetheless, both human and animal data suggest that a significant fraction of the general carcinogenic burden borne in the United States is due to insidious environmental exposure.

Chick Stroke

Many classes of cancer have viable candidates for causes, including radiation and specific pollutants (e.g., benzidine and tobacco smoking). However, the causes of heart disease and stroke have been elusive. Physicians doubt that circulatory diseases are caused by any unique, well-defined agents.

This last case involves a very potent toxin: (2, 3, 7, 8-dibenzoparadioxin), also known as dioxin. Figure 5 shows the chemical structure of this substance, which is shockingly toxic even in tiny quantities (parts per trillion of body weight).

In 1957, in the midwestern and the southeastern United States, there was a serious outbreak of chick edema disease. Hundreds of thousands

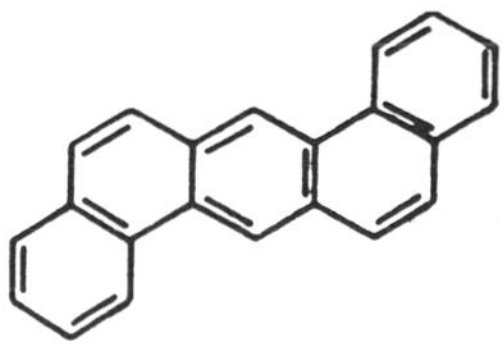

Dibenz [a,h] anthracene

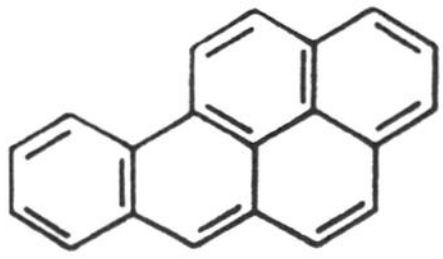

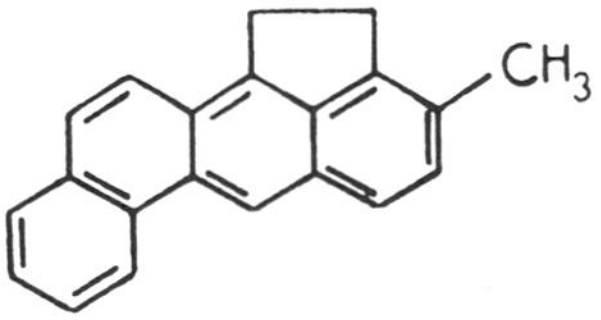

Benzo [a] pyrene   3-Methylcholanthrene

Figure 4
Three carcinogenic polycyclic hydrocarbons.

of chickens destined for supermarkets died rapidly and suddenly. In 1967 there was a second outbreak in the midwest. These chickens died with a pathology that had not been described in the literature, presumably as the results of a new disease. Chicken growers, who had lost millions of dollars, pressed for an aggressive investigation. The toxic agent was soon localized to tallow residues that feed companies had added to chicken feed to increase its caloric content at low cost.

Tallow is a heated and distilled lipidic material made, in part, from fat scraped off partially treated cow hides before they are sold to the leather industry. The hides had been dipped in a preservative called pentachlorophenol to prevent fungi and bacteria from growing on them during transport and storage. Pentachlorophenol, when heated under certain conditions, condenses to form 2,3,7,8-dibenzoparachlorodioxin. Thus, the feed additives contained low but toxic levels of dioxin containing compounds.

Humans may come into contact with dioxins in ways other than through the ingestion of contaminated chicken. Some fats obtained from tallow are used as additives in human foods. Some tallow by-

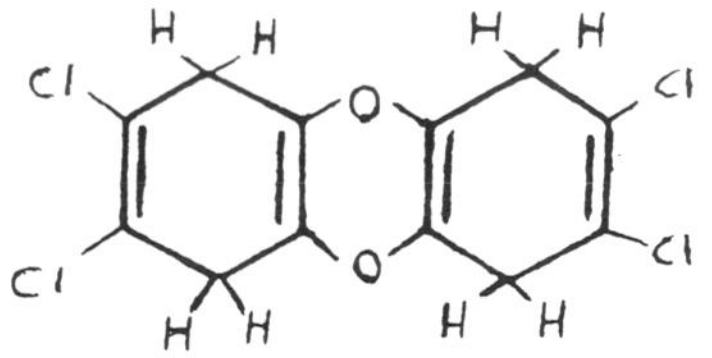

Figure 5
Dioxin.

products are used for industrial purposes. Cardboard and certain building materials are treated with preservatives that react to form dioxin upon combustion. How humans are affected by exposure to low levels of environmental dioxin is unknown, but the example provided by the poultry industry serves as a warning.

Conclusion

We can't assume that we humans or the plants and animals we depend upon are immune from environmentally induced diseases. In the long term, the solution is to include assessments of the biological and environmental affects of new agents in the initial decisions as to whether these substances should be produced and dispersed in the ecosystem. This effort is only beginning.

Readings

Hodge, L. 1977. *Environmental Pollution,* second edition. Holt, Rinehart and Winston.

King, J. 1997. Environmental Health. Interactive lecture tape, Environmental Evolution course, University of Massachusetts, Amherst.

Lippmann, M., and R. Schlesinger. 1979. *Chemical Contamination in the Human Environment.* Oxford University Press.

Moriarty, F. 1988. *The Study of Pollutants in Ecosystems.* Academic Press.

Nriagu, J. O. 1996. A history of global metal pollution. *Science* 272: 23–24.

Steingraber, S. 1997. *Living Downstream: An Ecologist Looks at Cancer and the Environment.* Perseus.

Appendix A
Teaching Strategy

Professional geologists, environmental scientists, and biologists can easily teach this material; in many cases they will be learning anew with their students. For the instructor, three books are indispensable supplements to this text: *Traces of Bygone Biospheres*, by Andrey Lapo, *Early Life*, by Lynn Margulis, and *A New Bacteriology*, by Sorin Sonea and Maurice Panniset. Appendix C lists introductory readings from many other fields. Both instructors and students can compensate for any deficiencies in background by consulting them. Plan to adapt our materials to your own needs and to the local ecology.

The course begins and ends at the planetary level. During the first three sessions the instructor leads an exemplary discussion about the geological time scale and the Gaia Hypothesis. Lynn Margulis's videotaped lecture "The Gaia Hypothesis and Early Life" (see below) is comparable to the material in this early portion of the course. The course becomes more participatory as the students grow increasingly familiar with the material, and culminates with students giving their own 10-minute presentations and leading class discussions.

To best appreciate this scientific activity, students should have at least four semesters of any combination of biology, chemistry, astronomy, physics, and geology. Like natural science itself, the program of study evolves and is infinitely expansible. Therefore we have listed in appendix C only the most highly recommended books. In addition to textbooks and secondary sources, it is essential that readings from the primary literature be assigned. Each semester we offer some ten to twenty primary sources as supplementary readings. Some we have used are listed in appendix C.

Students may be evaluated primarily by the quality of their participation and their oral presentations. Over the course of a semester, each student gives three 10-minute presentations and a 5-minute presentation on topics of his or her choice, based on the chapters in this volume. Guidelines for preparing these presentations are given below. We ask our students to list the titles of their talks on a large master sign-up calendar. This organizes the topics to be covered in a given class period and allows students to coordinate topics with their classmates. Several short written homework assignments are also required: the time and space assignments, a genetics worksheet, and a geology worksheet. Instructions and samples of these are included in this appendix and should be modified to suit the instructor and the students. In addition, our students create guides for field trips to local sites or institutions of biological interest. Field trips can include visits to museums of science, aquariums, Audubon Society nature centers, university collections, sewage-treatment plants, local conservation areas, and other places of ecological interest, such as salt marshes, bogs, and forests. Plans and questions for two sample field trips are given below. Each student is requested to provide general information about the accessibility of his field site before presenting sample questions and suggested answers. Field trip guides should include explicit directions, hours (where useful), and other information, as in the examples provided below.

Sample Schedule and Minimal Assignments

week 1	introductory session: interactive lectures and other resources
week 2	Gaia
week 3	Gaia
week 4	planetary background[1]
week 5	origin of life
week 6	early life and Archean record; Time Assignment due; agreement on plan for field trip

---

1. Possible visit sites for campus tour: computerized reference library services, map room, rock collection, geology department, petrographic thin section laboratory, meteorological laboratory, herbarium, greenhouse, electron microscope or microprobe laboratory, gas chromatographic-mass spectometry laboratory, etc. Check your campus for details.

week 7	Proterozoic eon and cell origins
week 8	symbiosis; Space Assignment due
week 9	Phanerozoic eon and animal origins; Genetics Worksheet due
week 10	plate tectonics and continental drift
week 11	pollution, history of environment
week 12	open discussion; Geology Worksheet and Field Trip Report due
week 13	oral final exam

Time Assignment

The purpose of the Time Assignment is to develop an appreciation for the vastness of geological time and for the remoteness of many events that affect us directly today. The assignment is discussed in the class during the fourth or fifth week and should be handed in two weeks later.

Each student is asked to think of and name an event of interest in the history of the cosmos, such as the origin of *Homo sapiens sapiens*, the origin of the Earth-Moon system, evidence of the first life on Earth, the origin of the universe, the orogeny of the Appalachian Mountains, or the end of the last glaciation. Each student must suggest one possibility; if there are sixteen students, the class members, by negotiation, must trim the list by agreeing to ten major events. The assignment is for each student to develop a written time scale (in any form) and place the events on it.

Our criteria for evaluation are the validity of the scale, the use of units (e.g., 1 centimeter = 500 million years), the accuracy of the placement of events on the scale, references, aesthetics, and originality. We present the finest contributions to the entire class and even, when they are extraordinary, post them in the classroom.

To follow up this assignment, we draw attention to Calder's *Timescale: An Atlas of the Fourth Dimension* and to the Elsevier poster of the geological time scale.

Space Assignment

The space assignment requires the students to integrate knowledge crossing various levels of linear dimension, from the molecular (e.g., Ångstroms) to the astronomical (Astronomical Units). Invariably, when asked to arrange the linear dimensions of objects on a scale, they recognize not only the need for a common unit, such as the meter, but also the need for an exponential scale of notation.

As with the Time Assignment, the exercise begins in class, and two weeks are given for completion. Each class member is requested to suggest an object relevant to the course whose linear dimension it would be useful to know. Typical examples are the diameter of the Earth, the distance across a hydrogen atom, the height of the biosphere from the abyss to the top of the troposphere, the diameter of a globular protein, the diameter of a coccoid bacterial cell, and the distance to the edge of the visible universe. After the ten entries of greatest interest to the class are agreed upon, each student pursues the assignment as she or he wishes. After the work has been submitted, we draw attention to Morrison and Morrison's book *Powers of Ten* and show the film of the same name.

Worksheets

Genetics and Mutation

(1) Starting with one cell (at generation 0) which divides at regular intervals, if no cells die, how many cells will there be after

1 generation? _____

2 generation? _____

3 generation? _____

4 generation? _____

5 generation? _____

10 generation? _____

20 generation? _____

N generation? _____

If you start with K cells in the above example, how many cells will there be after N generations?

(2) In March a particular pond has one water lily on it. Every two days the number of lilies doubles. On May 30 the entire pond is covered and looks like a giant lily pad. On what days was the pond one-eighth, one-quarter, one-half covered? Do you think water lilies have such a biotic potential? Define "biotic potential" and explain its significance. What is the point of this exercise for the study of evolution? (Hint: Define "natural selection" for yourself.)

(3) *Penicillium*[2] does not require biotin in its growth medium. *Neurospora*, a similar fungus, does require biotin. Is this observation consistent with the conclusion that biotin plays no role in the cellular biochemistry of *Penicillium*? Explain.

(4) Define and place each of the following on a temporal (functional) diagram that shows how protein synthesis works:

DNA

polysomes

mRNA

tRNA, GTP

rRNA

aminoacyl-tRNA synthetase

ribosomes.

(5) Give a real example (that is, specify an organism, a phenotype, and your source of information) for each of the following types of mutations:

point mutation (one or very few base changes)

deletion

chromosomal mutation

aneuploidy.

---

2. Genera, such as *Penicillium* and *Neurospora*, are capitalized and italicized; species names (specific epithets) are written in lower case and also italicized—for example, *Penicillium crysogenum*. Names of drugs, such as penicillin, are written without capitalization or special fonts.

(6) For each of the following, give a one-sentence definition and a referenced example:

replicon

transformation

transduction

transposition

conjugation.

(7) Of the following chromosomal mutations, which kind do you think is the most significant for evolution?

inversion

duplication

deletion

Explain your answer.

(8) Give a referenced example of the mode of action of a chemical mutagen and a measured value for mutation rate.

(9) In 1859 a gentleman farmer in southern Australia imported twelve pairs of rabbits from England to satisfy his passion for hunting. Within a few years, the rabbit became a major agricultural pest on the whole continent. Many ways of eradication were tried without success. In 1959, a virus producing myxomatosis, a disease fatal to rabbits, was deliberately released in the countryside and rapidly spread throughout Australia. It caused such a widespread epidemic that in some areas the rabbit census dropped by a factor of 100 in a matter of weeks. However, year after year, smaller percentages of the rabbits died. Explain what happened, using the concepts and terms of neo-Darwinism.

(10) Explain how artificial selection may have detrimental effects on organisms used in phenotypic breeding (for example, in breeding for coat color or for egg or milk production). (Hint: Look up artificial selection, phenotype-genotype relations, pleiotrophy, or genetic homeostasis in a comprehensive genetics text.)

(11) What is the "New Synthesis?" (Hint: Look up Julian Huxley or Ernst Mayr.)

(12) A woman with blood type Rh-negative O has four sons and four daughters with a man who has Rh-positive AB blood. Diagram the cross and show the expected blood types of their eight children. How is this example related to "evolution by natural selection"?

Geology

(1) Name and distinguish the three major types of rocks formed on Earth. What are the salient differences among them?

(2) List three localities in the Amherst area (western Massachusetts and the Connecticut River Valley) where you could expect to find a sample of each of the three general rock types. Specify these localities, indicating your references. Bring in (place it, labeled, in a plastic bag with your name, the rock type, and the locality) one identified rock sample from an identified local outcrop.

(3) We are familiar with elemental cycling at the global and local levels, but what is meant by the "rock cycle"? Draw a diagram of the rock cycle to illustrate your answer. (Hint: exogenic, endogenic).

(4) Briefly note the mechanisms behind orogeny (mountain-building), using the ideas of global plate tectonics. Give an example of an orogeny, including the relative movement of plates in the context of a specific geographical location.

(5) Cite two forms of field evidence (be specific about location) that have been used to establish the presence of former glacial activity in the Amherst area. When did the last glaciers recede from New England? (By "field evidence" we mean topographical structures and patterns, soil deposits, outcrops, and similar features visible in the landscape).

(6) How do meteorites help us to determine the age of the Earth?

(7) What is the fundamental difference between the lithification of detrital sediments (sandstones and shales) and the formation of carbonates?

(8) Define biomineralization. Name two examples of biogenic minerals. What is the evidence that these two minerals are biogenic? (Hint: Look at Lowenstam's book *On Biomineralization*.)

Suggestions to Students Preparing Class Presentations

Preparation

Read the assigned chapter and the relevant papers in your packet.

Make a list of topics or questions related to the chapter that intrigue you, including some which you think might interest your classmates.

Discuss your ideas with a former student or instructor. After you are sure your instructor or teaching fellow has accepted your topic, record and schedule your title on the master sign-up calendar.

Outline your presentation.

Presentation

Only one concept can be presented well in a 10-minute talk. Relate each point carefully to that concept. Use acetate diagrams, posters, slides, chalkboard, or whatever else is needed to make each point vivid and memorable. Occasionally repeat the main idea, indicating how the various ancillary points in the presentation are related to the concept. Talk slowly, and stop if you think that the class is not following a line of reasoning. At the end, summarize the entire talk in a few sentences like those at the beginning, and then invite questions from the class.

Know your material very well. Do not attempt to teach something you do not understand yourself. Discuss points that may be confusing to you with your classmates and the faculty before the talk.

Discuss your topic with any classmates who have signed up to speak on a similar topic. Coordinate the material to be presented; prevent redundancy by careful planning; divide aspects of the subject matter appropriately.

Discuss your presentation plan well in advance with colleagues (for example, graduate students or professors of zoology, geography, microbiology, or geology). They may be willing to lend you slides, rocks, or other materials. Borrow slides, if possible.

Make your own slides or transparencies. Check local resources for fossils, minerals, or other relevant illustrative materials.

Do not hesitate to consult introductory geology, astronomy, and biology textbooks on matters of background and terminology. Define carefully any new terms. If necessary, review introductory points with the class.

If your topic includes some controversial points, try to create a class discussion by asking provocative but clear questions.

Your presentation need not be formal or stiff. You may encourage class discussion or whatever you wish within your time allotment (usually ten minutes). The main requirements are that your presentation be informative, accurate, and interesting to your fellow students.

During your assigned presentation time, you alone are responsible for the quality of the class.

The Interactive Lecture Program

Essential to our way of teaching the Environmental Evolution course are the Interactive Lecture tapes, whereby each student, alone in a study carrel, listens to and queries scientists. The Interactive Lecture console, developed at MIT and Polaroid by Stewart Wilson, exemplifies our expectations for advanced science education in the 21st century.

It would, at present, be prohibitively expensive to duplicate the tapes, the electrowriter signals, and other materials. Since the entire program in the original interactive format is available at Boston University and at the University of Massachusetts at Amherst, potential teachers of environmental evolution are invited to visit or enroll in the course.

Our tapes are listed at the end of this appendix. The tape library, because of the importance of the scientists who participated in the recordings, is also an archive for scholars interested in the history of what Gerald Soffen calls "planetary evolutionary biogeochemistry." These taped materials, some of which are in a video format, are invaluable to the way we teach the course.

We use a seminar format based on the tape library, which is always expanding. The lecturers are eminent chemists, geologists, and biologists, most of whom are currently researching the early history of the Earth and the evolution of life. The contributing scientists are chosen not only for their direct involvement in the field of study but also for their ability to communicate clearly.

Each unit consists of the main portion of the lecture (always less than an hour long) and a comprehensive series of questions prepared by students and later answered by the lecturer. Supplementary visual aids (such as color slides, maps, and photographs) and a printed outline of the lecture accompany each tape. Most of the tapes include electrowriter signals that reproduce the lecturer's handwriting on paper.

The lecture format gives the student the opportunity to take part in an "active" learning process through simulated conversation. The student may selectively review the lecture material, obtain supplementary information, and clarify finer points by listening to the question-and-answer tapes. The interactive nature of the format makes it the student's responsibility to learn the material. The quantity of information each student acquires is largely a function of the time he or she puts into listening to, reviewing, and assimilating the material. Because students prepare for class by listening to the assigned audiotaped lecture, class time is free for debate, discussion, student presentation, and further development of the topics. As a result, a large amount of material is presented and understood in the limited class time of a single semester.

The Interactive Lecture console is housed in a comfortable atmosphere conducive to learning. Access to the lecture programs is available by appointment at the student's convenience. With additional course work, usually involving further study of "choice tapes" (also listed below), students may obtain graduate credit under the course title Environmental Evolution.

Tape Library[3]

"Extraterrestrial Organic Matter" (Willam Irvine, University of Massachusetts, Amherst)
Irvine discusses the presence and possible origins of interstellar and interplanetary organic matter. He describes the methods used to detect

---

3. The tapes are listed roughly in chronological order. Unless noted otherwise, each is approximately 45 minutes long.

such material and how some of it even lands on our planet. He suggests how extraterrestrial carbon compounds may even have influenced the evolution of life on Earth.

"Cosmochemical Evolution" (Cyril Ponnamperuma, Laboratory for Chemical Evolution, University of Maryland)
Ponnamperuma discusses the origin of life from interactions of organic chemicals on the prebiotic Earth and gives evidence supporting the chemo-evolutionary theory of the origin of life drawn from the NASA discoveries on meteorites and interstellar dust.

"Polymers before Monomers?" (Clifford Matthews, University of Illinois at Chicago; 15 minutes)
Matthews presents the evidence for his controversial hypothesis on the origin of proteins, contrasts his theory with that of Ponnamperuma, and describes its implications for observations of organic matter in meteorites and the outer planets.

"RNA vs. DNA" (video; Antonio Lazcano, Universidad Nacional Autónoma de México; 12 minutes)
Lazcano reviews various aspects of experimental origin-of-life research and argues that RNA preceded DNA as the genetic material of the earliest organisms.

"Evolution of the Atmosphere" (H. D. Holland, Harvard University)
Holland describes the first 500 million years of Earth's history, before there were any rocks on the planet. He explains how study of the rock record led him to believe that the composition of the atmosphere has changed dramatically since our planet's formation. In response to questions, he asserts that the Gaia hypothesis is not really necessary to help explain the past history of the planet's surface.

"Evidence of Earliest Life" (Paul Strother, Boston University)
Strother explains the criteria used for determining whether putative microfossils are actually biogenic, and examines the evidence for life in the early Archean eon based on thin sections of rocks from sites in Greenland, southern Africa, and northwest Australia.

"The Antiquity of Life" *(Elso Barghoorn, Harvard University)*
Barghoorn describes biota from three sites that elucidate the antiquity
of life: the Fig Tree Formation in the Swaziland System of southern
Africa, the Gunflint Iron Formation of North America, and the Bitter
Springs Formation of central Australia.

"Earliest Life: The Rock Record" *(video; Maud Walsh, Louisiana State University; 12 minutes)*
Walsh describes the geological sequence in the Swaziland System,
gives evidence that the oldest microfossils are found in what was once
a shallow marine environment, and discusses the geological processes
involved in the formation of mats and cherts.

"Life in the Proterozoic Eon" *(Andrew Knoll, Harvard University)*
Knoll describes the sedimentary rocks and some of the microfossils
preserved in them from 1 billion to 570 million years ago in the Draken
Formation of the Arctic Svalbard archipelago.

"The Ediacarans of the Late Proterozoic" *(Mark McMenamin, Mount Holyoke College)*
Life was abundant at sandy seashores in the late Proterozoic Eon, and
much of it was large. Strange three-sided repeat-patterned bodies,
spindle-shaped sunbathers, and bloated bag-like creatures abounded.
Most of these once-conspicuous organisms have no modern counter-
parts. McMenamin carefully rejects the hypothesis that these Ediacarans were animals. Certainly they were not plants. He outlines the
developmental pattern of several kinds of life just before the "Cambrian explosion." He recreates Rodinia (the pre-Pangean supercontinent) and its shoreline inhabitants from the abundance of fossil
records.

"Prokaryotic Motility, Eukaryotic Motility, and 'Rubberneckia'" *(Sidney Tamm, Boston University)*
In this set of three lectures, Tamm gives a thorough explanation of our
current understanding of the motility systems of prokaryotes and
eukaryotes, and of the unique motility system found in the de-
vescovinid protist "Rubberneckia."

"Symbiotic Theory: Cells as Microbial Communities" *(Lynn Margulis, University of Massachusetts, Amherst)*
Margulis presents a current view of the status of endosymbiotic theory. This theory postulates the origin of three classes of organelles of eukaryotic cells (plastids, mitochondria, and undulipodia) from separate lineages of bacteria.

"Spirochetes and the Origin of Undulipodia" *(Lynn Margulis, University of Massachusetts, Amherst)*
Margulis explains her hypothesis for the origin of undulipodia (eukaryotic "flagella" and cilia) within the endosymbiotic theory, and presents the status of experimental evidence supporting the hypothesis.

"Comparison of Planetary Atmospheres: Mars, Venus, and Earth" *(Michael McElroy, Harvard University)*
McElroy hypothesizes that the atmospheres of the inner planets were originally similar. He describes and compares the atmospheres of Mars, Venus, and Earth, and explores the mechanisms of planetary atmospheric evolution that may have occurred on each.

"Gaia" *(James Lovelock, F.R.S., Cornwall, United Kingdom)*
Lovelock suggests that the sum of the organisms on Earth forms a complex system that regulates its environment at the surface of the planet. Positing that the growth and the activities of organisms feed back on the environment in ways that ultimately affect the organisms, he presents evidence for the modulation of the atmosphere's chemistry and temperature. He argues the need to look at Gaia as an interacting, evolving whole.

"Life's Contribution to the Atmosphere" *(James Lovelock and Lynn Margulis)*
In part I, Lovelock and Margulis contrast the conventional and the Gaian views of the origin and history of Earth's atmosphere. In part II, they discuss evidence for and implications of life's control of atmospheric composition, temperature, acidity, and oxidation state.

"Continental Drift and Plate Tectonics" (Raymond Siever, Harvard University)
Siever describes the discoveries that led to the modern version of Wegener's continental-drift theory. He explains the processes of plate tectonics as revealed through studies of the distribution of fossils, the sea floor and its paleomagnetism, and the geography of volcanism, earthquakes, ridges, faults, and other geological features.

"Algal Mats of the Persian Gulf" (Stjepko Golubic, Boston University)
Golubic describes the different types of coastal microbial mats, communities that form the sabkha of Abu Dhabi. Today's layered sediment, with its ancient history, reveals the relationship of living communities to their potential preservation as both microfossils and laminated rocks (stromatolites).

"Stromatolites of Shark Bay, Australia" (Stjepko Golubic, Boston University)
Golubic describes the formation of different types of stromatolites forming today in saline basins. He discusses the relationship between the recent stromatolites of Shark Bay and the ancient laminated limestone rocks—stromatolites of Proterozoic and Paleozoic times—that were made by communities of microorganisms..

"Microbial Activity and Archean Gold Deposits" (Betsey Dyer, Wheaton College)
Dyer discusses the role that ancient bacteria may have played in forming the Archean gold deposits of Witwatersrand, South Africa. She reviews experiments that demonstrate the ability of modern bacteria to precipitate gold from solution and argues for their relevance to the reconstruction of past river-delta systems.

"The Microbial Community at Laguna Figueroa, Mexico" (John Stolz, Duquesne University)
Stolz discusses the stratified microbial community involved in the deposition of laminated sediments in Baja California, Mexico. He describes the site, the basic structure of a microbial mat, and the organisms that build and inhabit the mat.

"Biomineralization: Production of Minerals by Living Organisms" (Heinz Lowenstam, California Institute of Technology)
Lowenstam discusses some of the more than forty minerals produced biogenically inside cells or resulting from the activities of cells. The implications for the fossil record of biomineralization by selected animals (i.e., as teeth and skeletal materials) are detailed.

"Biodestruction and Stabilization of Mineral Surfaces" (Wolfgang Krumbein, University of Oldenburg)
The relationship of microbes to geological processes is sometimes destructive (as in the action of lichens on and in rock surfaces) and sometimes protective (as in the action of heterotrophs which produce a coating called "rock varnish"). The observation that a certain organism is present in a specific environment needs to be interpreted.

"Plant Chemical Signals and Phanerozoic Evolution" (Tony Swain, Boston University, and Robert Buchsbaum, Massachusetts Audubon Society)
Swain introduces the concept of "ecological hormones" (i.e., allelochemicals, semiochemicals) to encompass the interspecific chemical signals between plants and animals. He presents the chemical structures for major classes of these compounds (such as terpenoids and alkaloids). He relates the compounds by common metabolic pathways and ecological significance.

"The Genetic Mechanisms of Evolution" (Lynn Margulis, University of Massachusetts, Amherst)
Margulis discusses the classification of the diverse biota found on Earth and explains some of the basic evolutionary mechanisms that led to this great diversity of life.

"Origins of Life: Historical Development of Recent Theories" (Antonio Lazcano, Universidad Nacional Autónoma de México)
Lazcano presents a historical review of the major theories of the origins of life and discusses some of the twentieth-century experiments in origins-of-life research.

"Origins of Membranes: Structures and Functions" (David Deamer, University of California, Santa Cruz)
Deamer discusses the properties of contemporary liposome-forming lipid bilayers and presents his discovery of lipid-like material extracted from the Murchison meteorite. The presence of lipid-like material in meteorites provides a source on the early Earth of molecules that may have formed membranes of the earliest cells.

"The Theory of Plate Tectonics" (Raymond Siever, Harvard University)
Siever traces the development of ideas and the sequence of discoveries that led to the theory of plate tectonics, an all-encompassing theory that provides a basis for geological phenomena.

"Mammalian Evolution: Karyotypic Fission Theory" (Neil Todd, Boston University)
Todd explains the evidence that chromosomes have fissioned in the evolution of mammals. He explores mechanisms by which these genomic rearrangements may be passed through populations, enabling speciation events. He then uses his theory of karyotypic fission to explain the evolution of family lineages in various groups of tetrapods, such as pigs and dogs.

"Hallucinogenic Plants and Fungi of North America" (Richard Evans Schultes, Harvard University)
Schultes describes various ceremonial uses of hallucinogenic plants and fungi by native peoples of the North American continent. Prominent among his examples is psilocybin from the *Psilocybes* mushroom and mescaline from the peyote cactus.

"Hallucinogenic Plants and Fungi of South America" (Richard Evans Schultes, Harvard University)
Schultes describes the abundance of of hallucinogens, their origins and their uses, primarily in Amazonia.

"Plants, People, and Pollutants" (William Feder, Waltham Field Station, University of Massachusetts; 20 minutes)
Feder reviews the effects of atmosphere emissions of ozone and hydrocarbons on the physiology of economically important plants, and explores the extent to which the atmosphere links people and their agriculture.

"Environment and Disease" (Jonathan King, Massachusetts Institute of Technology)
In the only tape in this series concerned with the human environment, King points out that anthropogenic by-products of civilization dispersed through the environment have caused much human death and suffering. His three examples include heavy metal toxicity, colorant-induced bladder cancer, and preservative induced dioxin poisoning. He illustrates the general principal that the activities of organisms impact their environment in ways that change the health and the growth rate of the organisms' populations.

Supplementary Materials

"Burgess Shale" (videotape, 1992)
This CBC video explores the fossils of the Burgess Shale; hosted by David Suzuki.

"Five Kingdom Slide Set" (Margulis and Schwartz, 1989 and 1990)
Available from Ward's Natural Science Establishment, Inc., P.O. Box 92912, Rochester, NY 14692–9012.

"Five Kingdoms" (poster, 1991)

"Gaia: Goddess of the Earth" (videotape, 1985)
This video, produced for the BBC series *Horizons*, was shown in the United States as a *Nova* program.

"Gaia Hypothesis and Early Life" (videotape, 1984)

This lecture, given by Lynn Margulis at the NASA Lewis Research Center, is available on video cassettes from NASA Core, 15181 Route 58 South, Oberlin OH 44074.

"Geology of the Connecticut River Valley" (videotape, 1994)
This video, produced by Richard Little, Professor of Geology at Greenfield (Massachusetts) Community College, explains basic geological concepts as it explores the events that have occurred throughout the geologic history of the Connecticut River Valley.

"Geological Time Table" (poster, 1987)
This is a color poster compiled by B. U. Haq and F. W. B. Van Eysinga. The fourth revised, enlarged, and updated edition is available from Elsevier Scientific Publishing Company, Inc., 52 Vanderbilt Avenue, New York, NY 10017.

"Life on Ice" (videotape, 1989)
This video explains how the study of Antarctic microbial communities living at the bottom of ice-covered dry-valley lakes helps NASA researchers prepare for the search for the remains of ancient life on Mars.

"Natural Connections" (videotape, 1992)
This CBC video, hosted by David Suzuki, examines evolution, ranging in level from the cosmic to the cellular.

"Planet Earth: Fate of the Earth" (videotape, 1989)
This video examines the origin on life on Earth and basic geological principles.

"Powers of Ten" (film, 1978)
This classic and unparalleled 16-mm color film, which explores linear dimensions from the atomic scale to the universal scale, is available from Pyramid Films, P.O. Box 1048, Santa Monica, CA 90406.

"Ring of Truth: Clues" (videotape, 1987)
This *Nova* program, hosted by Philip Morrison, examines what happened to the Mediterranean Sea in the Miocene.

"Wonders of the Rain Forest" (videotape, 1990)
This superb film exploring the rainforest ecosystem was made by NHK TV Japan.

Appendix B
Five-Kingdom
Classification Scheme[1]

Superkingdom Prokaryota (Chromonemal Organization)

Kingdom Monera (Procaryotae)

Prokaryotic cells, bacteria. Nutrition absorptive (heterotrophic or autotrophic). Anaerobic, aerobic, facultatively anaerobic, microaerophilic or aerotolerant metabolism. Reproduction asexual and chromonemal; sex by conjugation with unidirectional recombination or mediated by small replicons (e.g., viruses, transposons, plasmids, transformation). Nonmotile or motile either by gliding or by bacterial flagella composed of flagellin proteins. Solitary unicellular, filamentous, colonial, or mycelial. Some produce sheaths, spores, or other multicellular resistant structures, sessile or stalked. Cell walls absent (Tenericutes), patchy (Mendosicutes), or composed of peptidoglycans between two lipoprotein membranes (Gracilicutes, Gram-negative) or external to the membrane (Firmicutes, Gram-positive).

SUBKINGDOM ARCHAEOBACTERIA

DIVISION: *Mendosicutes*

Phylum 1: Methanocreatrices: methane-synthesizing bacteria; anaerobic chemotrophs (*Methanobacterium*)

Phylum 2: Halophilic and thermoacidophilic bacteria: salt- and heat-tolerant bacteria (*Thermoplasma*)

---

1. See L. Margulis and K. V. Schwartz, *Five Kingdoms,* third edition (Freeman, 1998).

SUBKINGDOM EUBACTERIA

DIVISION: *Tenericutes*

Phylum 3: Aphragmabacteria: mycoplasmas, wall-less bacteria

DIVISION: *Gracilicutes*

Phylum 4: Spirochaetae: spirochetes (*Spirochaeta, Treponema, Cristispira*)

Phylum 5: Thiopneutes: anaerobic sulfate- or sulfur-reducing bacteria (*Desulfovibrio*)

Phylum 6: Anaerobic phototrophic bacteria: purple nonsulfur bacteria (*Rhodospirillum*), green sulfur bacteria (*Chloroflexus*), purple sulfur bacteria (*Chromatium*)

Phylum 7: Cyanobacteria: blue-green bacteria, blue-green algae (*Pleurocapsa, Nostoc, Oscillatoria*) and Chloroxybacteria: prokaryotic green algae (*Prochloron*)

Phylum 8: Nitrogen-fixing aerobic bacteria: (*Azotobacter, Rhizobium*)

Phylum 9: Pseudomonads: Gram-negative, aerobic heterotrophs (*Pseudomonas*)

Phylum 10: Omnibacteria: Gram-negative aerobic heterotrophic bacteria, enterobacteria, coliforms (*Escherichia, Salmonella*), prosthecate bacteria (*Caulobacteria*), acetic-acid bacteria (*Acetobacter*), Moraxella-Neisseria group (*Neisseria, Moraxella*), predatory bacteria (*Bdellovibrio*), microaerophilic bacteria (*Spirillum*), vibrios (*Photobacterium*), aerobic and facultatively anaerobic rods, chlamydias, and rickettsias.

Phylum 11: Chemoautotrophic bacteria: sulfur-oxidizing bacteria (*Thiobacillus*), ammonia-oxidizing bacteria (*Nitrobacter, Nitrosomonas*), iron-oxidizing bacteria (*Ferrobacillus*)

Phylum 12: Myxobacteria: heterotrophic aerobic gliding bacteria (*Beggiatoa*), fruiting myxobacteria (*Chrondromyces*)

DIVISION: *Firmicutes*

Phylum 13: Fermenting bacteria (*Clostridium*)

Phylum 14: Aeroendospora: aerobic endospore-forming bacteria (*Bacillus*)

Phylum 15: Micrococci: Gram-positive aerobes (*Paracoccus, Sarcina*)

Phylum 16: Actinobacteria: Gram-positive coryniform and mycelial bacteria (*Actinomyces, Streptomyces*)

Superkingdom Eukaryota (Chromosomal Organization)

Kingdom Protoctista

Eukaryotic cells: membrane-bounded nuclei invariably present, more than a single chromosome per cell. Nutrition heterotrophic, either ingestive or absorptive, or, if photoautotrophic, by measured photosynthetic plastids. All products of evolution of two or more integrated prokaryotic heterogenomic systems. Aquatic microorganisms exclusive of animals, plants, and fungi. Reproduction is asexual, premitotic, or eumitotic sexual. In eumitotic forms, meiosis and fertilization are present, but details of cytology, life cycle, and ploidy level vary from group to group. Organisms are solitary unicellular, syncitial (plasmodial, coenocytic) colonial unicellular, or multicellular. All lack embryos and complex cell junctions (e.g., desmosomes or septate junctions). Most bear undulipodia (eukaryotic flagella or their shorter homologues, the cilia) composed of microtubules in the [9(2) + 2] pattern. Species are aquatic. Primarily unicellular forms are sometimes called protists.

I. Subgroup of phyla in which members lack undulipodia at all stages; complex sexual cycles absent

Phylum 1: Rhizopoda: amebas, rhizopods

Phylum 2: Haplosporidia: haplosporidians, parasites

Phylum 3: Paramyxea: paramyxeans, parasites

Phylum 4: Myxozoa: myxozoans, fish parasites

Phylum 5: Microspora: microsporans, fish parasites

II. Subgroup of phyla in which members lack undulipodia; sexual cycles correlated with complex morphology present

Phylum 6: Acrasea: acrasids, cellular slime molds

Phylum 7: Dictyostelida: dictyostelids, cellular slime molds that form migrating "slugs" or "hats"

Phylum 8: Rhodophyta: rhodophytes, red algae, red seaweeds

Phylum 9: Conjugaphyta: gamophytes, conjugating green algae

III. Subgroup of phyla in which cells of members may reversibly form undulipodia; sexual cycles correlated with complex morphology absent

Phylum 10: Xenophyophora: xenophyophores, deep-sea macroscopic protists

Phylum 11: Cryptophyta: cryptomonads, some "phytoflagellates"

Phylum 12: Glaucocystophyta: glaucocystids, cyanelle-bearing algae

Phylum 13: Karyoblastea: karyoblastean giant amebas

Phylum 14: Zoomastigina: some "flagellates," some "parasites," amebomastigotes, bicoecids, choanomastigotes, diplomonads, pseudociliates, kinetoplastids, opalinids, proteromonads, parabasalians, retortamonads, pyrsonymphids

Phylum 15: Euglenida: euglenids, some "phytoflagellates"

Phylum 16: Chlorarachnida: chlorarachnids, "colored amebas"

Phylum 17: Prymnesiophyta: prymnesiophytes, some "phytoflagellates," algae

Phylum 18: Raphidophyta: raphidophytes, some "phytoflagellates," algae

Phylum 19: Eustigmatophyta: eustigmatophytes, some "phytoflagellates," eye-spot algae

Phylum 20: Actinopoda: acantharians, radiolarians, polycystinids, phaeodarians, heliozoans

Phylum 21: Hyphochytriomycota: some water molds, hypochytrids

Phylum 22: Labyrinthulomycota: slime nets, thraustochytrids

Phylum 23: Plasmodiophoromycota: plasmodiophorids, some "plant parasites"

IV. Subgroup of phyla in which cells of members may reversibly form undulipodia; meiotic-fertilization sexual cycles correlated with complex morphology present

Phylum 24: Dinomastigota: dinomastigotes, "dinoflagellates," some planktonic algae

Phylum 25: Chrysophyta: chrysophytes, golden-yellow algae

Phylum 26: Chytridiomycota: chytrids, monoblepharids, and other water molds

Phylum 27: Plasmodial slime molds: mycetozoa, "myxomycetes," acellular slime molds

Phylum 28: Ciliophora: ciliates, "infusoria," suctorians

Phylum 29: Granuloreticulosa: foraminifera and shell-less relatives

Phylum 30: Apicomplexa: apicomplexans, "sporozoan parasites"

Phylum 31: Bacillariophyta: diatoms, some algae

Phylum 32: Chlorophyta: green algae, pondweeds, green seaweeds

Phylum 33: Oomycota: oomycetes, some water molds, downy mildews

Phylum 34: Xanthophyta: xanthophytes, yellow-green algae

Phylum 35: Phaeophyta: phaeophytes, kelps, brown algae, brown seaweeds

Incertae Sedis

Phylum 36: Ebridians: some plankton

Phylum 37: Ellobiopsida: parasites

Kingdom Fungi

Haploid or dikaryotic cells; diploids undergo zygotic meiosis to form haploid spores. Organisms are filamentous (mycelial) or secondarily unicellular. They possess chitinous walls and always use absorptive nutrition. Cells always lack [9(2) + 2] undulipodia. The body plan, which may be branched, is composed of hyphae (coenocytic filaments that may be divided by perforate septa). Only single-cell forms are yeasts. Lack pinocytosis and phagocytosis. Extensive cytoplasmic streaming. Propagation by spores.

Phylum 1: Zygomycota: zygomycetes, molds (*Rhizopus, Mucor*)

Phylum 2: Ascomycota: sac fungi or ascomycetes, yeasts (*Saccharomyces*), molds (*Neurospora*)

Phylum 3: Basidiomycota: club fungi, rusts, smuts, mushrooms (*Agaricus, Coprinus*)

Phylum 4: Deuteromycota: fungi imperfecti (*Candida, Penicillium, Aspergillus*)

Phylum 5: Mycophycophyta: lichens, fungal component + cyanobacterial component, or fungal component + green algal component (*Cladonia, Xanthoria*)

Kingdom Animalia

Gametic meiosis; anisogamous fertilization; sperm and egg form a zygote, which cleaves to form diploid blastula; gastrulation and histogenesis generally follow to form multicellular adult with sex organs in which gametic meiosis occurs. Nutrition heterotrophic; sometimes ingestive by phagocytosis and pinocytosis, sometimes absorptive. Extensive cellular and tissue differentiation; desmosomes, septate junctions, gap junctions, and other differentiated connections between cells.

Subkingdom Parazoa

Phylum 1: Placozoa: *Trichoplax*

Phylum 2: Porifera: calcareous and siliceous sponges

Subkingdom Eumetazoa

Phylum 3: Cnidaria: coelenterates, hydroids, jellyfish, corals, sea anemones

Phylum 4: Ctenophora: comb jellies

Phylum 5: Mesozoa: mesozoans

Phylum 6: Platyhelminthes: flatworms, planarians, flukes, tapeworms

Phylum 7: Nemertina: nemertine worms

Phylum 8: Gnathostomulida: gnathostome worms

Phylum 9: Gastrotricha: gastrotrichs

Phylum 10: Rotifera: rotifers

Phylum 11: Kinorhyncha: kinorhynchs

Phylum 12: Loricifera: loriciferans

Phylum 13: Acanthocephala: spiny-headed worms

Phylum 14: Entoprocta: entoprocts or kamptozoa

Phylum 15: Nematoda: nematodes, roundworms (*Ascaris, Caenorhabditis*)

Phylum 16: Nematomorpha: horsehair worms

Phylum 17: Ectoprocta: bryozoa, moss animals

Phylum 18: Phoronida: phoronid worms

Phylum 19: Brachiopoda: brachiopods, lamp shells

Phylum 20: Mollusca: molluscs, monoplacophorans, solenogasters, chitons, toothshells, snails, bivalves, squids, octopuses, nautiloids

Phylum 21: Priapulida: priapulid worms

Phylum 22: Sipuncula: sipunculid worms, peanut worms

Phylum 23: Echiura: echiuroids, sea cucumbers

Phylum 24: Annelida: segmented worms, oligochete worms, polychaete worms, leeches, earthworms

Phylum 25: Tardigrada: tardigrades, water bears

Phylum 26: Pentastoma: pentastomes, tongueworms

Phylum 27: Onychophora: *Peripatus*

Phylum 28: Arthropoda: joint-footed animals, horseshoe crabs, sea spiders, scorpions, ticks, spiders, hornets

Phylum 29: Pogonophora: beard worms, tubeworms, vestiminiferans

Phylum 30: Echinodermata: echinoderms, sea lilies, starfish, brittle stars, sea urchins

Phylum 31: Chaetognatha: chaetognaths, arrow worms

Phylum 32: Hemichordata: acorn worms, *Rhabdopleura, Cephaladisas*

Phylum 33: Chordata: notochord-bearing animals, tunicates, seasquirts, ascidians, lancelets, *Amphioxus,* vertebrates, lampreys, hagfishes, cartilaginous fish, bony fish, amphibians, reptiles, birds, mammals

Kingdom Plantae

Multicellular organisms in which the haploid generation develops from spores, and fertilization produces the diploid embryo that develops into the mature sporophyte, which by meiosis produces spores. Photoautotrophic nutrition: chloroplasts contain chlorophylls *a* and *b*. Organisms exhibit advanced tissue differentiation, many lignified. Production of complex secondary compounds (e.g., polyphenolics, anthocyanins, alkaloids, terpenoids) is common.

DIVISION BRYOPHYTA: *nonvascular embryophytes*

Phylum 1: Bryophyta: hornworts, liverworts, mosses

DIVISION TRACHEOPHYTA: *vascular plants (xylem, phloem tissue)*

Phylum 2: Psilophyta: *Psilotum*

Phylum 3: Lycopodophyta: club mosses and quillworts (*Lycopodium, Selaginella, Isoetes*)

Phylum 4: Sphenophyta: horsetails (*Equisetum*)

Phylum 5: Filicinophyta: pteridophytes, polypodiophytes, ferns (*Polypodium, Osmunda*)

Phylum 6: Cycadophyta: cycads (*Zamia, Cycas*)

Phylum 7: Ginkgophyta: *Ginkgo*

Phylum 8: Coniferophyta: conifers, yews (*Taxus*), pine, spruce, fir (*Tsuga, Cedrus*)

Phylum 9: Gnetophyta: *Gnetum, Ephedra, Welwitschia*

Phylum 10: Angiospermophyta (Anthophyta, Magnoliophyta): flowering plants, monocots (grasses, orchids, lilies, palms), dicots (cactuses, roses, daisies)

Appendix C
Background Reading

General

Calder, N. 1983. *Timescale: An Atlas of the Fourth Dimension.* Viking.

Goldsmith, D., and T. Owen. 1992. *The Search for Life in the Universe.* Addison-Wesley.

Lapo, A. 1987. *Traces of Bygone Biospheres.* Synergetic Press.

Liebes, S., E. Sahtouris, and B. Swimme. 1998. *A Walk through Time.* Wiley.

Margulis, L., and K. V. Schwartz. 1998. *Five Kingdoms.* Freeman.

Morrison, R. 1999. *The Spirit in the Gene: Humanity's Proud Illusion and the Laws of Nature.* Cornell University Press.

Vernadsky, V. 1998. *The Biosphere,* ed. M. McMenamin. Copernicus.

Cells and Protists

Alberts, B. 1998. *Essential Cell Biology: An Introduction to the Molecular Biology of the Cell.* Garland.

Lipps, J. H. 1993. *Fossil Prokaryotes and Protists.* Blackwell.

Margulis, L., H. I. McKhann, and L. Olendzenski, eds. 1993. *Illustrated Handbook of Protoctista: Vocabulary of the Algae, Apicomplexam, Ciliates, Foraminifera, Microspora, Water Molds, Slime Molds, and the Other Protoctists.* Jones and Batlett.

Margulis, L., J. O. Corliss, M. Melkonian, and D. J. Chapman, eds. 2000. *Handbook of Protoctista: The Structure, Cultivation, Habitats and Life Histories of the Eukaryotic Microorganisms and Their Descendants Exclusive of Animals, Plants and Fungi.* Academic Press.

Chemistry

Atkins, P. 1998. *Physical Chemistry*. Freeman.

Hill, J. W., and D. K. Kolb. 1998. *Chemistry for Changing Times*. Prentice-Hall.

Jones, M. 1997. *Organic Chemistry*. Norton.

Oxtoby, D. W., W. A. Freeman and T. F. Block. 1998. *Chemistry, Science of Change*. Sanders.

Solomon, T. W. 1995. *Organic Chemistry*. Wiley.

Stryer, L. 1998. *Biochemistry*. Freeman.

Environment

Botkin, D. B., and E. A. Keller. 1998. *Environmental Studies: The Earth as a Living Planet*. Merrill.

Callenbach, E. 1998. *Ecology: A Pocket Guide*. University of California Press.

Dewdney, A. K. 1998. *Hungry Hollow: The Story of a Natural Place*. Copernicus.

Harbourne, G. 1993. *Introduction to Ecological Biochemistry*. Academic Press.

Holdrege, C. 1996. *Genetics and the Manipulation of Life: The Forgotten Factor of Context*. Lindisfarne.

Schaefer, V., and J. A. Day 1981. *A Field Guide to the Atmosphere*. Houghton Mifflin.

Gaia

Barlow, C. 1991. *From Gaia to Selfish Genes*. MIT Press.

Bunyard, P. ed. 1996 *Gaia in Action*. Floris.

Geology and Plate Tectonics

Briggs, J. C. 1987. *Biogeography and Plate Tectonics*. Elsevier.

Fortey, R. 1997. *Life: An Unauthorized Biography*. HarperCollins.

Gary, M., R. McAfee Jr., and C. L. Wolf. 1977. *Glossary of Geology*. American Geological Institute.

Seyfert, C. 1987. *The Encyclopedia of Structural Geology and Plate Tectonics*. Van Nostrand Reinhold.

Microbiology

Hott, J. G., ed. 1984—1989. *Bergey's Manual of Determinative Bacteriology* (four volumes). Williams and Wilkins.

Kristjansson, J. K. 1992. *Thermophilic Bacteria*. CRC Press.

Lederberg, J., ed. 2000. *Encyclopedia of Microbiology* (six volumes). Academic Press.

Madigan, M. T., J. M. Martinko, and J. Parker. 1996. *Brock Biology of Microorganisms*. Prentice-Hall.

Trueper, H. G., A. Balows, and H. G. Schlegel. 1988—1992. *The Prokaryotes* (four volumes). Copernicus.

Origin and Early Evolution of Life

Bengtson, S. 1994. *Early Life on Earth*. Columbia University Press.

Cloud, P. 1988. *Oasis in Space: Earth History from the Beginning*. Norton.

Schopf, J. W. 1999. *Cradle of Life*. Princeton University Press

Planetary Atmospheres

Grinspoon, D. H. 1997. *Venus Revealed: A New Look Below the Clouds of Our Mysterious Twin Planet*. Addison-Wesley.

Symbiosis

Margulis, L., and R. Fester. 1991. *Symbiosis as a Source of Evolutionary Innovation: Speciation and Morphogenesis*. MIT Press.

Sapp, J. 1994. *Evolution by Association*. Cambridge University Press.

Sapp, J. 1999. *What Is Natural?* Oxford University Press.

Appendix D
Geological Time[1]

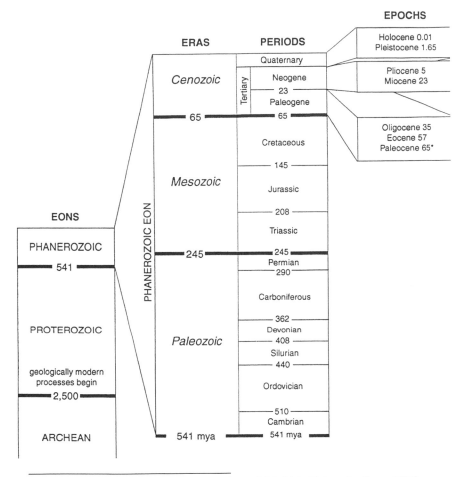

mya=millions of years ago
(not to scale)

EONS		ERAS	PERIODS	EPOCHS

EPOCHS

ERAS   PERIODS

Holocene 0.01
Pleistocene 1.65

Quaternary

Cenozoic   Neogene   Pliocene 5
                           Miocene 23
        — 23 —
        Paleogene

— 65 —   — 65 —

Oligocene 35
Cretaceous   Eocene 57
              Paleocene 65*

— 145 —

Mesozoic   Jurassic

— 208 —

Triassic

EONS

PHANEROZOIC   —245—   —245—
                       Permian
— 541 —   —290—

                       Carboniferous

                       — 362 —
PROTEROZOIC   Devonian
            Paleozoic — 408 —
                       Silurian
                       — 440 —
geologically modern
processes begin   Ordovician
— 2,500 —
                       — 510 —
                       Cambrian
ARCHEAN   541 mya   541 mya

Tertiary

PHANEROZOIC EON

---

1. Source: L. Margulis and D. Sagan, *What Is Life?* (Yale University Press, 1995).

Appendix E
Modes of Nutrition

The following table lists the sources of energy, electrons, and carbon for metabolism, giving examples of the growth of organisms to which the various prefixes apply.

Energy Sources	Electron sources (or hydrogen donors)	Carbon sources	Organisms and their hydrogen or electron donors

Photo- (light)	Litho- (inorganic and C_1 compounds)	Auto- (CO_2)	Prokaryotes
			  Chlorobiaceae, H_2S, S
			  Chromatiaceae, H_2S, S
			  Rhodospirillaceae, H_2
			  Cyanobacteria, H_2O
			  Chloroxybacteria, H_2O
			
			Protoctista (algae) H_2O
			Plants, H_2O

		Hetero- $(CH_2O)_n$	None
	Organo- (organic compounds)	Auto-	None

		Hetero-	Prokaryotes
			  Chromatiaceae, org. comp.[1]
			  Chloroflexaceae, org. comp.[1]
			
			  Heliobacteriaceae, org. comp.[1]
			  Rhodomicrobium, C_2, C_3

Energy Sources	Electron sources (or hydrogen donors)	Carbon sources	Organisms and their hydrogen or electron donors

Chemo- (chemical compounds)	Litho-	Auto-	Prokaryotes
			  methanogens, H_2
			  hydrogen oxidizers, H_2
			  methylotrophs, CH_4,
			    $CHOH$, etc.
			  ammonia, nitrite oxidizers,
			    NH_3, NO_2^-

		Hetero-	Prokaryotes
			  "sulfur bacteria," S
			  manganese oxidizers, Mn^{++}
			  iron bacteria, Fe^{++}
			  sulfide oxidizers,
			    e.g., *Beggiatoa*
			  sulfate reducers
			    e.g., *Desulfovibrio*

	Organo-	Auto-	Prokaryotes
			  clostridia, etc., grown
			  on CO_2 as sole source
			  of carbon (H_2,-CH_2)

		Hetero-	Prokaryotes (most)
			  (including nitrate,
			  sulfate, oxygen and
			  phosphate[2] as terminal
			  electron acceptors)
			Protoctista[3] (most)
			Fungi[3]
			Plants[3] (achlorophyllous)
			Animals[3]

Table devised in collaboration with R. Guerrero, from *Handbook of Protoctista*, 1990.

1. Organic compounds (e.g., acetate, proprionate, pyruvate).
2. Detection of phosphine: I. Dévai, L. Felföldy, I. Wittner, and S. Plósz. 1988. New aspects of the phosphorus cycle in the hydrosphere. *Nature* 333: 343–345.
3. Oxygen as terminal electron acceptor.

Appendix F
Field Trips

We recommend that students be required to complete at least one field trip in this course. It may be sufficient for younger, less experienced students simply to fill out an existing field-trip form. Advanced students may use an existing form as a model to develop their own field trips. All trips are designed to take no longer than an hour once the student arrives at the site.

Two sample field trips and a list of other trips available in the area of Amherst, Massachusetts are provided here. We hope that instructors will generate comparable lists of field trips in their local areas.

Hitchcock Center for the Environment (Larch Hill Conservation Area)

Hours: Wednesday through Saturday, 9 A.M.–4 P.M.

Directions: From the center of Amherst, follow South Pleasant St. (Route 116) south one mile. Turn right at the sign reading "Larch Hill and Hitchcock Center." Admission is free.

The Hitchcock Center for the Environment is an environmental education facility. Many classes, lectures, workshops, and special programs are offered there each year. The Hitchcock Center has a knowledgeable staff and a resource center that houses an extensive collection of books, maps, posters, tapes, and other environmental education materials.

The Larch Hill Conservation Area consists of 27 acres of moist forests, swampland, and ponds (including a vernal pool) with several well-maintained trails and a wheelchair-accessible boardwalk. Larch Hill's two forested areas are connected by a narrow path crossing through private property. Larch Hill is owned by the Town of Amherst and operated under the jurisdiction of the Amherst Conservation Commission.

Before walking the trails, familiarize yourself with the Hitchcock Center environmental education building, including the upstairs resource room. Note the many types of information and teaching tools available here, including the special collection of books, maps, and other information about the Connecticut River Watershed.

Questions

(1) List one upcoming event, lecture, or class at the Hitchcock Center that may be relevant to the subject matter of this class.

(2) To whom or what does the name Hitchcock refer, and why is the name well suited for a nature center in this area?

Now you may begin the outside portion of this field trip. Follow the trails marked on the map available inside the Resource Center, beginning at the wooden boardwalk. Walk along the boardwalk and mentally note the types of plants, animals, and other organisms you see and the conditions of the habitats in which they are found. This field trip will take you through areas of both the near and the far part of the Larch Hill Conservation Area.

(3) The name Larch Hill refers to the resident European larch trees (*Larix decidua*). This is a deciduous conifer, unlike many other conifers which do not lose their needles in the fall and are known as evergreens. These trees grow in swampy areas and serve as indicators of the wet and acidic conditions that persist in these environments. Name at least one other plant you see that is an "indicator" of swampy and/or acidic conditions.

(4) Look at the trees around you. What type(s) are they (e.g., deciduous, evergreen, mixed)?

(5) Notice the forest floor and its vegetation. By early May the ground below the trees is covered with low-lying plants. Would you expect to

find such extensive ground cover in a forest composed mostly of evergreens? Why or why not?

(6) Do you see any red maple trees (*Acer rubrum*)? If you are taking this field trip in early spring, you may notice dangling red flowers on the leafless red maples. If you are taking this field trip in late spring, the maples may have already shed their flowers and may be recognizable by their characteristic leaves. Maples are wind-pollinated trees, and like many other wind-pollinated trees maples flower in early spring. This is in contrast to many insect-pollinated trees, such as apple trees, which flower later in the year, after their leaves have fully developed. What might be the advantages of the alternative timings of wind-pollinated and insect-pollinated trees?

(7) Larches are not the only type of conifer found on Larch Hill. Another conifer common to Larch Hill is the eastern white pine (*Pinus strobus*). It is possible to estimate the age of a white pine by counting the rings of branches, or knots where branches once existed, along the length of the tree. Find a white pine of average size relative to other white pines in the area and estimate its age.

(8) At the large black dot on the map (where the trails meet before the crossover into the far wooded area), stop and note the very large white pine tree. The presence of this very large, many-trunked white pine with intact and well-developed (huge) lower branches might be evidence that this area was not always heavily wooded. Why is this? What other characteristics of the surrounding area or neighboring trees might also support this hypothesis?

(9) As you cross through the clearing to the second wooded area, note one or more species of plant that is unique or limited to this open habitat. Why do you think these plants do not appear in the woods?

(10) After crossing into the second wooded area, stop for a moment and look around you. How do these woods differ from the previous woods (i.e., do you see different types of trees, or differences in the amount or type of forest-floor vegetation; do you see or hear different birds or animals)?

(11) Note the poison ivy (*Toxicodendorn radicans*) growing on or among the trees. Does poison ivy possess simple leaves arranged in threes, or are its leaves compound and made of three leaflets each? (Hint: A true leaf can be distinguished by the presence of an axial bud in the angle of intersection of the leaf's petiole with the stalk.)

(12) Among the trees are some spiny-looking bushes with elongate red berries and alternate leaves. These are *Berberis thungergii*, or Japanese barberry, an insect-pollinated species introduced from Japan. Were the berries on the bush produced this year, or last year? How do you know?

(13) You may notice some small plants with three-section leaves and green and white pin-striped pitcher-shaped structures. These plants are called Jack-in-the-pulpit (*Arisaema atrorubens*). The pitcher-like structure, called a *spathe*, houses a spike structure called a *spadix*, which is the site on which the plant bears its tiny flowers and later its fruit. How are these plants pollinated? (Hint: Are there any stagnant-water-loving, blood-sucking animals swarming about you as you try to write down your answers?)

(14) When you reach the pond, study it carefully. How is the water entering the pond? Is a pond a lotic or a lentic environment? Is this pond oligotrophic, eutrophic, or dystrophic? Explain your answer.

(15) Continue until you reach the vernal pool. How does the water enter this pool? Why are vernal pools so important to the life cycles of many amphibians?

(16) You may notice some sphagnum moss on the ground in the vicinity of the vernal pool. What is the role of sphagnum moss in pond succession (the gradual transformation of a pond environment into a land environment)?

(17) As you continue your walk, look for other mosses. Where do you see mosses most frequently (on the ground, on dead wood, on live trees, etc.)? Do mosses seem to be more abundant in wetter or drier areas? Might this have to do with the life cycle of this organism? Explain.

(18) Have you seen many fungi during this field trip? Have you seen any lichens? What is the difference between a fungus and a lichen? What kinds of weather are best for finding mushrooms? List at least one place where you have seen a fungus.

(19) List two examples of symbioses that you have seen during your walk.

The formal part of this field trip is now over. You may now conduct further explorations on your own or return to the Hitchcock Center via the trails of your choice.

Readings

Hitchcock Center. Welcome to the Hitchcock Center and the Larch Hill Conservation Area (pamphlet); Newsletter & Program Guide.

Lincoff, G. H. 1994. *National Audubon Society Field Guide to North American Mushrooms.* Knopf.

Palmer, E. L. 1975. *Field Book of Natural History.* McGraw-Hill.

Peterson, L. A. 1977. *A Field Guide to Edible Wild Plants of Eastern and Central North America.* Houghton Mifflin.

Richardson, J. 1981. *Wild Edible Plants of New England: A Field Guide.* DeLorme.

Wernert, S. J., ed. 1982. *North American Wildlife: An Illustrate Guide to 2,000 Plants and Animals.* Reader's Digest Association.

Pratt Museum of Natural History, Amherst College

Hours: 9 A.M.–3:30 P.M. weekdays, 10 A.M.–4 P.M. Saturdays, noon–5 P.M. Sundays

Directions: Enter the Amherst College campus at the main entrance on Route 116 and follow the loop around to the far end. Parking is available along the loop. The museum is at the far end of the loop, next to the Merrill Science Center. Admission is free.

The Pratt Museum of Natural History houses outstanding collections and exhibits of vertebrate and invertebrate paleontology, minerals, and crystals. Specimens collected since the 1830s from all over the world have been assembled to display the evolution and ecology of major groups of animals as well as the geological processes that have formed the Earth and its many unique structures.

This tour consists of three sites.

Site 1 is the corridor. Enter the Pratt Museum and walk to your right until the stairs are behind you.

Site 2 is in the center of room B, which contains a three-dimensional geomorphological model of the surface rock of the Connecticut River Valley.

Site 3 is the main section of the museum. Straight ahead and to the left of the entrance is room A. While answering the questions at this site, which involves observing the large mammal articulated skeletons of the main exhibit, you will find it extremely helpful to climb the staircase, turn right, and enter room C. While you are in the upper gallery, note the modern and fossil shells from Florida and the superb ammonite cephalopods.

Questions

Site 1

(1) Look up at the enormous jaws of the largest fish that ever lived. What is its scientific name?

(2) In the jaw of this fish, how many sets of replacement teeth can you count?

(3) When did this fish live? Name the geological "time-rock" division and indicate how many million years ago this was.

(4) What are stromatolites? Sketch the stromatolitic limestone and label its surface.

(5) How many stromatolite heads can you count? Where are these stromatolites from?

Site 2

(6) Continue toward room B and the exhibit. Look at the large three-dimensional model of the valley. On the model, find the location at which you live. What color is the representation of the surface rocks where you live?

(7) Using the same color key and the color key in the displays around the model, find the age of the rocks that come to the surface. What type of rock are they (sedimentary, igneous, or metamorphic)? Explain your answer by describing the surface geomorphology of your local home location.

(8) Go around the corner to the cabinet that displays the dinosaur fossils that have been found in the Connecticut River Valley. The fact that *Coelophysis* lived here permits us to infer conditions at the time these animals roamed. Describe the local environment at the time large populations of *Coelophysis* lived.

(9) What evidence have we that *Coelophysis* engaged in cannibalism?

Site 3

(10) In the center of room A are three huge skeletons. Classify these skeletons from kingdom to species.

(11) Two of the animals are extinct. Which one is not?

(12) "Mastodon" means "breast teeth." The cups of these teeth seem to be rounded, unlike those of the other two large skulls. Which of the three are most closely related?

(13) Notice the large tusks. Are they inset in the upper or lower jaws? Do all have two tusks, or can you see any that have more than two? Please explain.

The major exercise at site 3 is to compare the three skeletons in order to reconstruct the lives of these three animals. To make the comparison, observe the skeletons from above as well as looking at their labels.

(14) What were their geographical and stratigraphic ranges?

(15) How do they differ in size?

(16) What are possible differences in their feeding habits?

(17) One of the fossils lived in more forested environments and another in more steppe-like or tundra open country. Speculate on the habitat of each and give evidence for your hypothesis.

Go to the bird exhibit (in room A) and look at the wall cabinet in which the fossil *Archaeopteryx* and a related dinosaur are compared.

(18) List two similarities and two differences between *Archaeopteryx* and the small dinosaur.

(19) When did *Archaeopteryx* live?

(20) Now look at the modifications of the birds' bills and sketch four of the different types of bills of birds modeled here. Indicate in what way the bills are correlated with the types of habitats in which the birds dwell (for example, sketch bills of fruit-eating, seed-eating, fish-spiking, algal-mud-eating, nut-eating or nectar-feeding birds).

Other Field Trips in the Amherst Area

Amethyst Brook Conservation Area

Amherst Wastewater Treatment Facility

Cider Mill Pond

Dinosaur Footprints at Holyoke Heritage State Park

Durfee Conservatory

Greenfield Community College Rock Garden

Harvard Forest

Quabbin Reservoir

Rattlesnake Gutter

Robert Bartlett Fishway at Holyoke Dam

Appendix G
The Evolution of Life, from Its Mysterious Origins More Than 3 Billion Years Ago to the Present

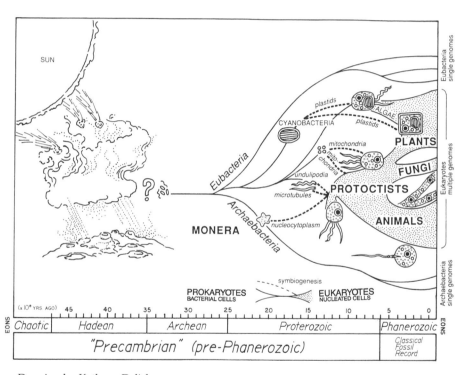

Drawing by Kathryn Delisle

Glossary

acritarch A hollow, microscopic, roughly spherical organic walled fossil removed from silicate rock with hydrofluic acid.

actin A class of proteins that are major constituents of the 7-nanometer-wide microfilaments of eukaryotic cells. Actin microfilaments have molecular weights of approximately 44,000; they are involved in the contraction of muscles and many other intracellular movements.

actinolite A bright green or grayish green silicate material of the amphibole group, $Ca_2(Mg,Fe)_5SiP5.5MJ68O_{22}(OH)_{MJ62}$, which may contain manganese. Actinolite, a type of asbestos, occurs in long, slender, needle-like crystals or in fibrous, radiated, or columnar forms in metamorphic rocks (such as schists) and in altered igneous rocks.

active margin Continental margin characterized by earthquakes, igneous activity, or mountain uplift–i.e., tectonic activity resulting from convergent or transform plate motion.

adaptive radiation The evolution from generalized, primitive species to diverse, specialized species, each adapted to a distinct mode of life. See also *cladistic episode* and *speciation*.

adiabatic (lapse rate) In thermodynamics, this term pertains to the relationship between pressure and volume when a gas or fluid is compressed or expanded without either giving or receiving heat. In an adiabatic process, compression causes a rise in temperature; expansion causes a drop in temperature.

aerobiosis The metabolism of an organism (an *aerobe*) that is active and capable of completing its life cycle only in the presence of gaseous oxygen (O_2). The oxidative breakdown of food molecules and the derivation of energy from them, in which the terminal electron acceptor is O_2, is known as *aerobic respiration*. A zone or an environment in which gaseous O_2 is present is called *oxic*.

aerosol (atmospheric particulate) A *sol* in which the dispersion medium is a gas (usually air) and the dispersed or colloidal phase consists of solid particles or liquid droplets; e.g., mist, haze, most smokes, and some fogs.

albedo The relative reflectivity of a body compared with that of a perfectly diffusing surface, measured on a scale from 0 to 1. Black surfaces reflect no incident light and have an albedo of 0. White surfaces reflect all incident light and have an albedo of 1.

algae A diverse group of eukaryotic, oxygenic, aquatic, photosynthetic, protoctists, including single-cell or few-celled forms (protists) and many multicellular descendants (e.g., rhodophytes, phaeophytes, and other seaweeds).

algal laminate See *stromatolite*.

algal pillar See *stromatolite*.

allelochemical See *semiochemical*.

anaerobiosis The metabolism of an organism capable of completing its life cycle in the absence of gaseous oxygen (O_2). An environment or a zone from which gaseous O_2 is absent, either because of physical exclusion or because of the activities of organisms capable of removing it, is called *anoxic*.

anoxygenic photosynthesis Photosynthesis in which H_2 or H_2S is the hydrogen donor and no O_2 is produced.

antibody A protein produced by lymphoid cells (plasma cells) in response to chemicals or microbial substances (antigens) and capable of interacting specifically with its antigen (or a chemically similar substance), leading to the amelioration, removal, or alteration of that antigen from the animal.

antigen A substance that, upon introduction into the body of vertebrates and some marine animals, stimulates the production of specific antibodies. An antigenic protein molecule may carry several distinct sites, epitopes, or antigenic determinants.

archaeobacteria (Archaea) A distinct group of prokaryotes, as determined by their ribosomal RNA, lipids, and other properties, which includes the methanogenic, the extreme halophilic, and certain acidophilic, thermophilic sulfur bacteria.

asthenosphere Upper portion of the mantle; the layer of the Earth just below the *lithosphere*, composed of rocks more plastic than the surface, in which isostatic adjustments take place, magmas may be generated, and *seismic waves* are strongly attenuated. It is equivalent to the upper mantle of the Earth.

atmosphere The mixture of gases that surround a planet. On Earth it is chiefly oxygen (20 percent) and nitrogen (79 percent), with some argon (1 percent), some carbon dioxide (0.03 percent), and smaller quantities of hydrogen, helium, methane, krypton, neon, nitrogen oxides, xenon, and other gases.

autopoiesis Organismal self-maintenance; metabolically active, self-bounded, and self-generating system (e.g., a cell). A prerequisite to reproduction and thought to have preceded reproduction in evolution; also spelled *autopoesy* or *autopoiesy*. From the Greek for "self-making."

banded iron formation A distinct type of sedimentary rock consisting of alternating layers of more and less oxidized iron oxides embedded in a chert matrix. Most of the economically important concentrations of iron in the world are found in Proterozoic (2.5 billion to 570 million years ago) banded iron formations.

basalt Dark-colored, fine-grained igneous rocks (iron, calcium, magnesium silicates), whether intrusive or extrusive.

biocoenosis See *community*.

biogenesis The production of a living cell, a mineral (e.g., a $CaCO_3$ shell), a gas (atmospheric oxygen), or a structure (stromatolite) from a parent organism or a living community of organisms. The doctrine that all life has been derived from previously living organisms.

biomineralization The formation of minerals by living organisms. Two kinds are known: *biologically controlled* or *matrix-mediated* biomineralization, i.e., intracellular precipitation of a given mineral type under genetic control of the cell (magnetite in magnetotactic bacteria, calcite by *Coleps* or coccolithophorids) and *biologically induced* biomineralization, i.e., production of acid, which changes local pH, or other environmental alterations that in turn cause potentially mineralizable material to precipitate (e.g., extracellular precipitation of iron and manganese oxides by *Leptothrix*, *Bacillus*, or other bacteria; precipitation of amorphous calcium in lakes due to algal activity).

biosphere The place where all the living things on Earth (the biota) reside. Extending from the top of the troposphere to below the abyss, it is the environment of the system of life at the surface of the Earth.

biota The sum of the living matter (all organisms) on Earth (the flora, fauna, and microbiota taken together).

biotic potential The number of organisms that can be produced in a single generation, or unit of time, which is characteristic of the species, measured in maximum number of offspring per generation, maximum number of spores produced per year, or equivalent terms. It illustrates the tendencies of organisms to increase exponentially when their conditions for material growth are satisfied.

bioturbation The disruption, churning, stirring, or other movement and local disturbance of sand, mud, or other sediments by live organisms.

calcification The hardening of tissue in live material or the replacement of organic material by calcium salts (especially $CaCO_3$) in fossilization.

calcium carbonate The common rock-forming mineral CaCO₃ (limestone).

carbonaceous chondrite A type of stony *meteorite* containing organic matter formed non-biologically.

chasmolith An ecological term referring to microorganisms living in rock crevices produced by erosion or by endolithic organisms. See *endolith* and *epilith*.

chert A hard, extremely dense or compact, dull to semi-glassy, cryptocrystalline sedimentary rock, consisting dominantly of cryptocrystalline silica with lesser amounts of micro- or cryptocrystalline quartz and amorphous silica (opal); it sometimes contains impurities such as calcite, iron oxide, and the remains of siliceous and other organisms.

cladistic episode An adaptive radiation in evolution, i.e., the splitting of a lineage of descent into two species or higher taxa.

clast An individual constituent, grain, or fragment of a sediment or rock produced by the mechanical weathering (disintegration) of a larger rock mass from the supporting, protective structures of animals, plants, or microbes, whether whole or fragmentary.

coacervate An aggregation of colloidal droplets that form when a solution of polypeptides, nucleic acids, and polysaccharides is shaken. Interpreted by A. I. Oparin to be a type of "protobiont."

coenocyte Plasmodium; syncitium. A multinucleate structure (thallus) lacking septa or cell walls; thallus with siphonous, syncitial, or plasmodial organization.

community A unit in nature composed of populations of organisms of different species living in the same place at the same time. *Microbial communities* are those lacking significant populations of animals and plants. A group of species living together as a community is also known as a *biocoenosis*.

crust See *lithosphere*.

cybernetic system An engineered, regulatory, multi-component control complex with sensory, amplification, and positive and negative feedback properties.

Daisyworld A first attempt by James Lovelock to apply cybernetic-style modeling to the Gaia hypothesis to demonstrate how the surface temperatures of terrestrial planets might be modulated by biota. The Daisyworld biota consists of light and dark daisies whose differential growth rates in varying temperature regimes determine changes in planetary albedo.

diagenesis (mineral) Recombination or rearrangement of a mineral that results in a new mineral.

diagenesis (sedimentary) All the chemical, physical, and biological changes, modifications, or transformations undergone by a sediment after its initial deposition (i.e., after it has reached its final resting place in the current cycle of erosion, transportation, and deposition) and during and after its lithification, exclusive of any surficial alteration (weathering) and metamorphism. *Early diagenesis* refers to diagenesis occurring immediately after deposition or burial. *Late diagenesis* refers to deep-seated diagenesis which occurs a long time after deposition, when sediment is more or less compacted into a rock, but still in the realm of pressure-temperature conditions similar to those of deposition.

dolomite   Calcium-magnesium carbonate, a common rock-forming rhombo-hedral mineral, $CaMg(CO_3)_2$, found in extensive beds as a compact limestone or dolomite rock; it is also precipitated directly from seawater, possibly under warm, shallow conditions. See *limestone*.

eclogite   High-temperature-and-pressure equivalent of basalt from Earth's interior. This bimineralic mantle rock forms the matrix for diamonds explosively exuded from kimberlite pipes. It is composed of garnite and clinopyroxene (an iron-poor calcium, magnesium silicate) and occasional traces of diamond, graphite, corundum, rutile (titanium dioxide), or coesite (high-temperature-and-pressure form of quartz).

ecological hormones   See *semiochemicals*.

ecosystem   A unit in nature composed of communities or organisms in which the biologically important elements (carbon, sulfur, nitrogen, phosphorus, oxygen, etc.) entirely cycle within the unit. The biologically essential chemical elements tend to cycle more rapidly within ecosystems than between them.

Ediacaran biota   A group of centimeter- to meter-sized creatures thought by many to represent an unusual, extinct phylum of life, a sixth kingdom to be added to the existing five (plants, animals, fungi, protoctists, and bacteria).

elongation factor TU (EFTU)   A protein that complexes with ribosomes to promote elongation of polypeptide chains; it dissociates from the ribosome when translation is terminated. Elongation factor TU is responsible for alignment of the AA-tRNA complex in the "A" site of the ribosome in protein synthesis.

endolith   An ecological term describing microorganisms living in tiny openings in rocks or rock crevices that have been produced by the metabolic activities of the endolithic organisms themselves. See *epilith* and *chasmolith*.

epilith   An ecological term referring to the biota living on the surface of rocks and/or stony material. See *chasmolith* and *endolith*.

ESA   European Space Agency (France, Germany, Netherlands, Italy).

eubacteria   All bacteria other than the Archaeobacteria (or Archaea) (i.e., mycoplasms, omnibacteria or Gram-negative rods, myxobacteria, cyanobacteria, actinobacteria, etc.). They differ from the archaeobacteria in that their cell walls contain neuraminic acid, and they have distinctive lipids, tRNAs, rRNAs, and RNA polymerases.

eukaryote A nucleated organism (protoctist, fungus, animal, or plant).

eukaryotic "flagella" See *Undulipodium.*

evaporite A nonclastic sedimentary rock composed primarily of minerals produced from a saline solution that became concentrated by the evaporation of the solvent; especially, a deposit of salt precipitated from a restricted or enclosed body of seawater or from the water of a salt lake. An example is gypsum, a widely distributed mineral consisting of hydrous calcium sulfate: $CaSO_4F2H_2O$. Gypsum is the most common sulfate mineral and is frequently associated with *halite* and anhydrite in evaporites or in thick, extensive beds interstratified with limestones, shales, and clays (especially in rocks of Permian and Triassic age).

facies Layer of rocks; the unit of study in stratigraphy. The sum of all primary lithologic and paleontologic characteristics exhibited by a sedimentary rock and from which its origin and environment of formation may be inferred; the general aspect, nature, or appearance of a sedimentary rock produced under or affected by similar conditions; a distinctive group of characteristics that differs from other groups within a stratigraphic unit.

Faint Young Sun Paradox Since the Archean eon, solar luminosity has increased as determined by all models of stellar evolution based on modern cosmogeny, and therefore the surface temperature of the Earth should have been below freezing in the past or boiling now; yet fossil evidence suggests that the Earth's surface temperature has remained constant during that time (from 3.4 billion years ago to the present), or that it has decreased.

fermentation Any process in which energy derived from metabolism (catabolism) of organic substrates is used in the generation of ATP via substrate-level phosphorylation. In all fermentations, the degradation of organic compounds in the absence of gaseous oxygen yields energy, while other organic compounds act as the terminal electron acceptors. Bacteria capable of obtaining energy via fermentation are known as fermenting bacteria.

flagellum An extracellular structure of some bacteria composed of homogeneous protein polymers, members of a class of proteins called flagellins; moves by rotation at the base; relatively rigid rod driven by a rotary motor embedded in the cell membrane that is intrinsically nonmotile and sometimes sheathed. See *undulipodium.*

formation A geomorphological unit of study in field geology; a geographically distinguishable naturally formed topographic feature, commonly differing conspicuously from adjacent objects or material, or being noteworthy for some other reason; especially a striking erosional form on the land surface.

fossil Any remains, trace, or imprint of a plant, animal, or microbe, or communities formed by them, that has been preserved, by natural processes, in the Earth's crust since some past geologic time; any evidence of past life. It is termed a *microfossil* if it is too small to study without the aid of a microscope, whether it is the remains of a microscopic organism or part of a larger organism. A sedimentary structure which consists of a fossilized track, trail, burrow, tube, boring, or tunnel resulting from the life activities (other than growth) of an organism, made on or in soft sediment at the time of its accumulation, is termed a *trace fossil*.

Gaia hypothesis The idea that the biota regulates specific aspects of the biosphere, that life on Earth forms a single metabolic physiological system in which more than 30 million types of organisms metabolize, grow, and die, each producing and removing gas. Each interacts with the elements C, H, O, N, P, and S. Their interactions lead to modulation of the Earth's temperature, acidity, and atmospheric composition. The idea was first stated in the late 1960s by the atmospheric chemist James Lovelock.

Galileo NASA space mission (1990–2000) to orbit Jupiter and send a probe to contact and analyze the Jovian atmosphere.

gene The unit of study in analysis of heredity of all organisms. From mating organisms and study of the distribution of traits in their offspring it can be inferred that hereditary units (genes) occupy specific positions (loci) within the nuclear genome, or chromosome in eukaryotes. Composed of DNA, these units of function can show one or more specific effects on the phenotype of the organism; they can mutate to one or more allelic forms or recombine with other such units. Three classes of genes are recognized: (1) *structural genes*, which are transcribed into mRNA and then translated into polypeptide chains, (2) *structural genes*, which are transcribed into rRNA and tRNA molecules which are used directly, and (3) *regulatory genes*, which are transcribed but serve as recognition sites for enzymes and other proteins involved in DNA replication and transcription. See *genome* and *replicon*.

genome All the genes carried by an individual or cell, i.e., a single gamete (or haploid organism); the bacterial genophore (nucleoid) and its plasmids or, in diploid eukaryotes, the set of chromosome pairs. The minimal sum of the genetic material required to determine an organism or set of genes inside an organelle. See *gene* and *replicon*.

geochemical process Chemical changes occurring in rocks (i.e., organic compound transformation during diagenesis in the formation of shale or coal from mud, or chemical transformation of one mineral to another under the pressure and temperature changes in metamorphism.)

geochronology The study of time in relation to the history of the Earth, especially by the absolute-age and relative dating systems developed for the purpose. See *radioactive decay*.

Glossopteris **flora** The Gondwanaland plant community dominated by this genus of cycadofilicalean (seed fern) trees, indicating the southern-hemisphere distribution of Paleozoic forests.

gypsum An evaporite mineral consisting of hydrous calcium sulfate ($CaSO_4F2H_2O$). See *evaporite.*

halite The evaporite mineral NaCl. It is native salt, occurring in massive, granular, compact, or cubic-crystalline forms and having a distinctive salty taste.

hematite The common iron mineral -Fe_2O_3. It is found in igneous, sedimentary, and metamorphic rocks, both as a primary constituent and as an alteration product. See *banded iron formation.*

hydrolysis The splitting of a molecule into two or more smaller molecules with the addition of the elements of H_2O. See *polymer.*

Hypersea Terrestrial ecosystems in which fungi connect tree roots and deliver nutrient salts to members of the contiguous communities.

isotope One of two or more species of the same chemical element having the same number of protons in the nucleus but different atomic weights, (i.e., different numbers of neutrons).

isotopic fractionation The relative enrichment of one isotope over another in a system, due mainly to the differential effects of temperature but also to kinetic effects, activity coefficients, etc. on the slight mass differences of the isotopes.

karst A type of topography formed over limestone, dolomite, or gypsum by dissolving or solution, that is characterized by closed depressions or sinkholes, caves, and underground drainage.

limestone A sedimentary rock consisting chiefly (more than 50 percent by weight or by areal percentages under the microscope) of calcium carbonate, primarily in the form of the mineral calcite, and with or without magnesium carbonate; specifically, a carbonate sedimentary rock containing more than 95 percent calcite and less than 5 percent dolomite.

Lipalian lithification The conversion of newly deposited, unconsolidated sediment into a coherent and solid rock, involving processes such as cementation, compaction, desiccation, crystallization, recrystallization, and compression. It may occur concurrent with, or shortly or long after, deposition.

lithosphere The solid portion of the Earth at its surface, above the mantle, as compared with the *atmosphere* and the hydrosphere; the *crust* of the Earth. See *asthenosphere.*

Magellan NASA space mission (1990–1992) to analyze from orbit the surface geomorphology of Venus, using radar.

magnesium calcite A variety of calcite: $(Ca,Mg)CO_3$. It consists of randomly substituted magnesium carbonate in a lattice of calcite. See *limestone.*

magnetite A black, isometric, strongly magnetic, opaque mineral of the spinel group: $(Fe,Mg)Fe_2O_4$. It constitutes an important ore of iron. See *banded iron formation*.

mantle See *asthenosphere*.

Mariner U.S. flyby missions (1964–1971) to Mars and other inner planets.

metabolism The sum of enzyme-mediated biochemical reactions that continually occur in cells and organisms and provide the material basis for autopoiesis.

meteorite Any meteoroid that has fallen to the Earth's surface in one piece or in fragments without being completely vaporized by intense frictional heating during its passage through the atmosphere. Most meteorites are believed to be fragments of asteroids and to consist of primitive solid matter similar to that from which the Earth was originally formed. There are three classes of meteorites: *Stony meteorites* consist largely or entirely of silicate minerals and comprise more than 90 percent of all meteorites seen to fall. *Iron meteorites* consist generally of nickeliferous iron (solid solution of iron with 4 percent to 30 percent or more of nickel). *Carbonaceous chondrites* are friable, dull black, chondritic stony meteorites containing hydrated, clay-type silicate minerals (usually fine-grained serpentine or chlorite) and considerable amounts and a great variety of organic compounds (hydrocarbons, fatty and aromatic acids, porphyrins).

microbe Microorganism. Prokaryote, fungus, or protoctist requiring a microscope for visualization.

microbial community A community made up of microorganisms.

microbial mat A benthic structure composed of a community of microorganisms, usually dominated by phototrophic bacteria, such as the cyanobacterium *Microcoleus,* that bind and trap sediment (and sometimes actively precipitate minerals). Microbial mats are living precursors of stromatolites.

microfossil See *fossil*.

microtubule A slender, hollow structure made primarily of tubulin proteins (alpha-tubulin and beta-tubulin), each with a molecular weight of about 50,000, arranged in a heterodimer. Microtubules are of varying lengths but usually invariant in diameter at 24–25 nanometers. They form the substructures of axopods, mitotic spindles, kinetosomes, undulipodia, haptonemata, nerve-cell processes, and many other intracellular structures. See *mitotic spindle* and *undulipodia*.

mitotic spindles Composed of microtubules, kinetochores, and often centrioles or centrosomes. Transient proteinaceous structures associated with mitotic cell division limited to eukaryotic organisms.

monomer Subunit of a polymer.

monophyly The condition of a trait or a group of organisms that evolved directly from a common ancestor. Sister taxa are monophyletic. See *taxon*.

mutagen A chemical causing alteration in the DNA.

NASA The National Aeronautics and Space Administration of the United States.

negative feedback mechanism A property of a system in which production or amplification of a material or a product leads to inhibition of that same production. For example, in feedback inhibition of metabolic control, the end product of a metabolic pathway acts as an inhibitor of an enzyme, or step, within that pathway.

oncogenes Genes that when damaged lead to cell transformation. Damage to these genes causes cancer.

oolite A sedimentary rock, usually a limestone, made up chiefly of ooliths cemented together. An *oolith* is one of the small, round, accretionary bodies in a sedimentary rock resembling the roe of fish and having diameters of 0.25–2 mm (commonly 0.5–1 mm).

organelle "Little organ"; a visibly distinct structure inside any type of cell, composed of a complex of macromolecules and small molecules. Examples include those lacking genomes (i.e., carboxysomes, ribosomes) or containing their own genomes (mitochondria, plastids, and nuclei).

organic geochemistry A science within geology or chemistry that studies naturally occurring carbonaceous and biologically derived substances of geological interest.

organo-sedimentary structures Biogenic rocks such as stromatolites.

oxic environment (zone) See *aerobiosis*.

oxidation The combination of a molecule with gaseous or atom oxygen or the removal of hydrogen from a molecule. Since electrons are transferred to the oxidizing reagent, which becomes reduced, oxidation and reduction (q.v.) are always coupled in what are called oxidation-reduction reactions. Reduction is classically defined as the addition of hydrogen or electrons. Natural forms of oxidized and reduced sulfur are shown in the diagram on the next page.

oxidation state The propensity of a compound to accept electrons (or their equivalent, H atoms). Oxidation states vary from fully oxidized (oxygen itself, great tendency to violently accept electrons) to hydrogen gas (nonoxidized, fully reduced, no tendency to accept more electrons). The change of oxidation state of elements such as sulfur or nitrogen (from fully oxidized sulfate through fully reduced nonoxidized sulfide). Intermediates such as thiosulfate or elemental sulfur are crucial for chemical transformations in nature. Life is based on incessant changes in oxidation state of carbon, sulfur, hydrogen, and nitrogen.

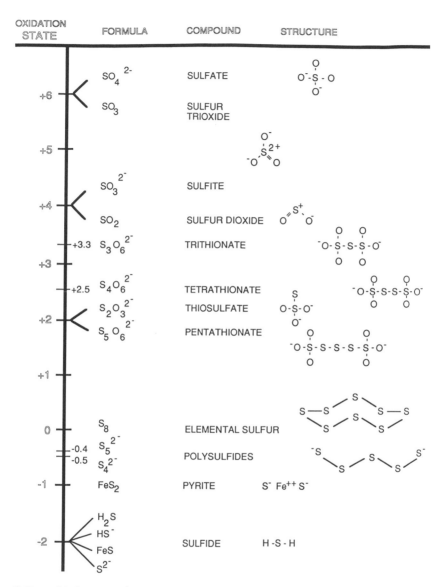

Sulfur oxidation states in nature.

oxidizing atmosphere An atmosphere that contains oxidized gases such as CO_2, H_2O, and N_2, and detectable (greater than trace) amounts of gaseous, free oxygen (O_2).

oxygen The eighth element of the periodic table. It has an atomic number of 8, a valence of –2, and a molecular weight of 16 grams per mole. See *oxidation state*.

ozone O_3, triatomic oxygen. A highly reactive gas with both tropospheric and stratospheric sources.

paleomagnetism The study of natural remanent magnetism in order to determine the intensity and direction of the Earth's magnetic field in the geologic past.

passive margin Leading border or subducting margin of tectonic plate, i.e., in the East Pacific, to be compared with trailing margin (i.e., mid-Atlantic rift), at which new igneous materials are forming. Continental margin characterized by thick, relatively undeformed sediments with only limited tectonism related to divergent plate motion

petrographic thin section Polished slices of rock thin enough to allow light to pass through them; used to detect microfossils in a cryptocrystalline matrix. See *chert*.

pheromone An example of a semiochemical.

photo-oxidation The production of singlet-state oxygen ($^1O_2$), a very powerful oxidant and rapidly lethal upon formation in the cell, by the reaction of certain photosensitive pigments with light in the presence of O_2 (molecular oxygen).

photosynthesis A mode of nutrition that permits cell growth with light captured by chlorophyll as energy source, usually accompanied by the production or organic matter from carbon dioxide and a hydrogen donor (such as hydrogen gas, H_2; water, H_2O; hydrogen sulfide, H_2S). In *oxygenic photosynthesis*, H_2O is the hydrogen donor and O_2 is produced. An organism capable of growth and metabolism using only light energy and inorganic carbon reduction (from CO_2 to all needed organic constituents of its cells) is a *photoautotroph*. See *anoxygenic photosynthesis*.

phototroph An organism that fills its energy requirements from light. See *photosynthesis*.

phylogeny Family tree. Diagram or other representation of hypothesized sequences of ancestor/descendant relationships of groups of organisms reconstructed from hypotheses of their evolutionary history.

Pioneer NASA missions to Venus, Jupiter, and Saturn (early 1970s through mid 1980s). Goals included reconnaissance, elucidation of structure, dynamics, and chemical composition of the atmosphere and ionosphere, and study of magnetic fields and solar-wind interaction.

plasmodium See *coenocyte*.

plasmogenesis Concept of the origins of protoplasm from inorganic or pre-biotic colloidal materials. This term refers mainly to an early-twentieth-century movement dedicated to the experimental study of the origins of life, domi-nated by Alfonso Herrera.

plate tectonics Global tectonics based on an Earth model characterized by a small number (10–25) of large, broad, thick plates (blocks composed of both continental and oceanic crust, and mantle). Continents are the raised portions of the plates. Each plate "floats" on some viscous underlayer in the mantle and moves more or less independently of the others and grinds against them (like ice floes in a river), with much of the dynamic activity concentrated along the periphery of the plates. Plates are propelled from the rear by *sea-floor spreading;* oceanic crust is increased by convective upwelling of magma along the mid-oceanic ridges. The older crust moves away from the new material at a rate of 1–10 cm per year. This is thought to provide the power source for plate tectonics. Linear margins of plates are locations of volcanic, earthquake, and other tectonic activity.

polar wandering curve The apparent movement during geologic time of the Earth's rotation and magnetic poles, suggested by shifts in the climatic zones and by paleomagnetic determinations. Possibly all indications of polar wan-dering can be accounted for by continental displacement. See *plate tectonics.*

polymer A macromolecule composed of a covalently bonded collection of repeating subunits or monomers linked together during a repetitive series of similar chemical reactions. Each strand of DNA is a linear polymer of nucleo-tide monomers. Characteristic structure of proteins (amino acid monomers), nucleic acids (nucleotide monomers), and many other substances.

population A local (geographically defined) group of nonspecific organisms sharing a common gene pool; also called a *deme*. A group of organisms belonging to the same species and living in the same place at the same time.

prebiotic chemistry Laboratory experiments or theoretical calculations of chemical interactions thought to have occurred on the Hadean or Archean Earth prior to the appearance of life but relevant to it.

primary metabolism The metabolism involved in the production and utiliza-tion of primary metabolites, those components required for cells, i.e., minimal autopoietic units.

primary metabolite An organic compound that is produced metabolically and is essential for completion of the life cycle of the organism that produces it (e.g., any of the 20 protein amino acids or nucleotides in RNA and DNA). Chemical component required for autopoiesis.

prokaryote An organism or a cell having bacterial structure and lacking membrane-bounded nuclei.

protein synthesis The process of protein formation by cells. The formation of a protein directed by a specific messenger RNA (mRNA) molecule is known as *translation*. The formation of an RNA molecule (including mRNA) from a DNA template by complementary base pairing is called *transcription*. This process is mediated by an enzyme known as *RNA polymerase*. Translation occurs in a *ribosome* (q.v.). (Ribosomes consist of two unequal subunits bound together by magnesium ions. These subunits consist of roughly equal parts of ribosomal RNAs (rRNA) and protein.) As the 5′ end of the mRNA moves through the ribosome, a lengthening polypeptide chain is produced. *Transfer RNAs* (tRNA) are molecules that function here to transfer amino acid residues to the growing polypeptide chain. The newly formed protein is released once the 3′ end of the mRNA has moved through the ribosome.

proteinoid microsphere Spheres which are products of thermal copolymerization of amino acids. Ranging in diameter from 0.1 to 10 m, they can be seen with a light microscope.

protoctists Algae, amebas, ciliates, slime molds, foraminiferans, seaweeds, water molds, and other eukaryotes; all multigenomic aquatic eukaryotes except plants, animals, or fungi. *Protists* are the few-celled or single-celled members of the kingdom.

pseudomorph A mineral whose outward crystal form is that of another mineral species; it has developed by alteration, substitution, incrustation, or paramorphism.

punctuated equilibria A term describing a pattern seen in the fossil record of relatively brief episodes of speciation followed by long periods of species stability. See *speciation*.

radioactive decay The spontaneous radioactive transformation of one nuclide to another, or of the energy state of the same nuclide (a species of atom characterized by the number of neutrons and protons in its nucleus). Essential to geochronology, this process is quantified in radiometric dating methods used to determine the age, in years, of rocks.

redox potential A quantifiable measure of the relative susceptibility of a substance to oxidation and reduction. See *oxidation* and *oxidation state*.

reduction See *oxidation*.

regolith A general term for the entire layer of fragmental, loose, incoherent or unconsolidated rock material, of whatever origin (residual or transported) and of very varied character, that nearly everywhere forms the surface of the land and overlies or covers the more coherent bedrock. Found on the surfaces of the moon, Mars, and Earth, it includes rock debris (weathered in place) of all kinds: volcanic ash, glacial drift, alluvium, loess and eolian deposits. On Earth, detritus, plant remains, and soil are included.

regression The retreat or contraction of the sea from land areas, and the consequent evidence of such withdrawal. The opposite of *transgression*.

.

replication A duplicating process requiring copying from a template; i.e., the copying of genetic material.

replicon A genetic element that behaves as an autonomous unit during DNA replication. In bacteria, the genophore (visible in electron microscopy as the nucleoid) functions as a single large replicon, whereas eukaryotic chromosomes contain hundreds of replicons in series. Small replicons include plasmids, viral nucleic acids, transposons, or other DNA capable of autonomous replication. See *gene* and *genome*.

reproduction Any process that augments the number of cells or organisms; not to be confused with sex or replication. See *sex, replication*.

respiration The oxidative breakdown of food molecules and release of energy from them; the terminal electron acceptor is inorganic and may be O_2 or, in anaerobic organisms, nitrate, sulfate, or nitrite. See *oxidation*.

reverse transcription DNA synthesis from an RNA template, mediated by reverse transcriptase. See *protein synthesis*.

ribosome See *protein synthesis*.

RNA polymerase See *protein synthesis*.

rRNA See *protein synthesis*.

sabkha A salt flat or low salt-encrusted plain restricted to a coastal area, as on the Arabian peninsula along the Persian Gulf. See *microbial mat*.

scytonemine Brownish, light-protective sheath pigment of unknown chemical composition found in cyanobacteria such as members of the genus *Scytonema*.

secondary metabolism The metabolic pathways that produce and utilize secondary metabolites. See *secondary metabolite* and *primary metabolite*.

secondary metabolite A metabolically produced organic compound that is not essential for completion of the life cycle of the organism that produces it (e.g., alkaloids, flavonoids, and tannins). They seem to play primarily ecological roles, and may serve as pheromones or stress-response compounds (phytoalexins). See *secondary metabolism* and *semiochemicals*.

secondary plant metabolites See *semiochemicals*.

sedimentary rock A rock resulting from the consolidation of loose sediment that has accumulated in layers; e.g., a *clastic rock* consisting of mechanically formed fragments of older rock transported from its source and deposited in water, air, or ice, or a chemical rock (such as rock salt or gypsum) formed by precipitation from solution, or a biogenic rock (such as certain limestones) consisting of the remains or secretions of plants, animals, and microbes.

sedimentary sequence (stratigraphic sequence) A set of deposited beds that describes the geologic history of a region.

seismic wave A general term for all elastic waves produced by earthquakes or generated artificially by explosions.

semiochemicals Chemical signals released in the environment that mediate interactions between different species. Examples are *allelochemicals* (chemicals secreted by an organism that influence behavior in a member of another species), which are the chemicals introduced into the environment by one species to suppress the growth or reproduction of another. Some allelochemicals are *secondary plant metabolites*. A chemical exchanged between members of the same species that affects behavior is termed a *pheromone*. Examples of such pheromones are the sex attractants, alarm substances, aggregation-promotion substances, territorial markers, and trail substances of insects. Generally these signals may be termed *ecological hormones*.

serial endosymbiosis theory The theory that undulipodia, mitochondria, and plastids originated respectively as motile, respiring, and photosynthetic free-living bacteria that established symbioses with other bacterial hosts, such as the extant *Thermoplasma*–i.e., that these organelles began as xenosomes. The theory that eukaryotic cells evolved from bacterial ancestors by a series of symbiotic associations that occurred in a specific temporal sequence.

sex Any process that recombines genes (DNA) in an individual cell or organism from more than a single parental source. Sex may occur at the nucleic acid, nuclear, cytoplasmic, and other levels.

silica The chemically resistant dioxide of silicon: SiO_2. It occurs naturally in five crystalline polymorphs: in cryptocrystalline form, in amorphous and hydrated forms, in less pure forms (i.e., sand and chert), and combined in silicates as an essential constituent of many minerals.

siliceous rock A rock containing abundant silica, especially free silica rather than silicates.

silicification A process of fossilization whereby the original organic components of an organism are replaced by silica, as either quartz, chalcedony, or opal.

sinter A chemical sedimentary rock deposited as a hard incrustation on rocks or on the ground by precipitation from hot or cold mineral waters of springs, lakes, or streams; specifically, siliceous sinter (an opaline variety of *silica* deposited as an incrustation by precipitation from the hot mineral waters of a hot springs or geyser) and calcareous sinter.

soil Organic-rich *regolith* of planet Earth.

solar luminosity The total amount of radiant energy at all wavelengths emitted into space per unit time by the entire solar surface. See *Faint Young Sun Paradox*.

speciation 1. The evolutionary process leading to the division of an ancestral species into offspring species that coexist in time; horizontal evolution or speciation; cladogenesis. 2. The gradual transformation of one species into another without an increase in species number at any time within the lineage; vertical evolution or speciation; phyletic evolution or speciation. See *sympatric speciation* and *phylogeny*.

Stardust A NASA probe intended to collect the first comet sample from deep space. Launched in 1999, it will contact Comet Wild in 2004 and return to Earth in 2006.

stratigraphy Representation, in geology of rocks, in which layers or strata are arranged as to geographic position and chronological order of sequence.

stratosphere An upper portion of a planetary atmosphere, above the troposphere and below the mesosphere, characterized by relative uniform temperatures and horizontal winds. On Earth, its lower limit varies from about 8 to 20 km; its upper limit is around 45 km. The temperature in this region is around $-75°C$.

stromatolite Laminated carbonate or silicate rocks, organo-sedimentary structures produced by growth, metabolism, trapping, binding, and/or precipitating of sediment by communities of microorganisms, principally cyanobacteria. Produced by benthic bacterial communities, they are still called *algal pillars* and *algal laminates*.

subtidal zone See *tidal zone*.

supratidal zone See *tidal zone*.

symbiogenesis Evolutionary innovation by establishment of permanent symbioses. As originally coined by K. S. Merezhkovsky in 1920, the term is defined as the origin of organisms (species) through combination and unification of two or many organisms entering into symbiosis.

symbiont See *symbiosis*.

symbiosis The prolonged physical association of two or more organisms belonging to different species. The levels of partner integration in a symbiosis may be behavioral, metabolic, gene product, or genic. A *symbiont* is a member of a symbiosis; an organism that lives with another of a distinct species or kind for most of the life cycle of both. The permanent symbiosis is referred to as the *holobiont* (e.g., the lichen); the partners (algae, fungi) are the *bionts*.

sympatric speciation An uncommon process by which populations inhabiting (at least in part) the same geographic range become reproductively isolated until they form new species. See *speciation*.

syncitium See *coenocyte*.

tannin (tannic acids) Secondary metabolites from the bark and fruit of many plants which are derivatives of flavonoids (condensed tannins) or triesters of glucose or other sugars (hydrolyzable tannins) with one or more trihydroxy-benzene carboxylic acid. Several types of polyphenolic organic compounds.

taxon (plural: taxa) The general term for a taxonomic unit, whatever its rank. Examples range from most inclusive (kingdom, phylum, class) to least inclusive (genus, species, variety).

taxonomic unit A named group of organisms, that are placed in the taxon on the basis of features they have in common, or on the basis of their ancestry, or both.

terrestrial planet A planet similar to Earth in terms of size and mean density and possession of derived oxidizing atmospheres. Usually the inner planets, Mercury, Venus, Earth, and Mars, but sometimes Pluto is included.

tidal zone (also littoral zone) Pertaining to the benthic ocean environment or depth zone between high water and low water; also, pertaining to the organisms of that environment. *Subtidal zone* refers to that part of the littoral zone that is between low tide and about 100 meters. *Supratidal zone* pertains to the shore area marginal to the littoral zone, just above high-tide level.

transcription See *protein synthesis.*

transform fault A strike-slip fault characteristic of midoceanic ridges and along which the ridges are offset. Analysis of transform faults is based on the concept of *sea-floor spreading.* See *plate tectonics.*

transgression The spread or extension of the sea over land areas, and the consequent evidence of that advance (such as strata deposited unconformably on older rocks, especially where new marine deposits are spread far and wide over the former land surface). Also, any change (such as rise of sea level or subsidence of land) that brings offshore, typically deep-water environments to areas formerly occupied by nearshore, typically shallow-water conditions, or that shifts the boundary between marine and nonmarine deposition, or between deposition and erosion outward from a marine basin. The opposite of *regression.*

translation See *protein synthesis.*

transposon One kind of transposable element in both prokaryotes and eukaryotes that is immediately flanked by inverted repeat sequences, which in turn are immediately flanked by direct repeat sequences. Transposons usually possess genes in addition to those needed for their insertion (e.g., genes for resistance to antibiotics, sugar fermentation, etc.).

tRNA See *protein synthesis.*

tropopause See *troposphere.*

troposphere The lowest level of a planetary atmosphere, in which the temperature decreases steadily with increased altitude. On Earth extending from the surface of its upper boundary, the *tropopause* (q.v.), at a height of about 8 to 20 km, depending on the latitude and the time of year. Turbulence is greatest in this region, and most of the visible phenomena associated with the weather occurs here (for example, cloud formation). See *stratosphere.*

tubulin Major protein of undulipodia and mitotic spindles. See *microtubule.*

unconsolidated sediment In geology, clastic, unlithified material (precursor to conglomerate) in which consolidation resulting from deposition was too rapid to give time for complete settling. See *sedimentary rock.*

undulipodium A cell-membrane-covered motility organelle usually showing feeding or sensory functions and composed of at least 200 proteins. Microtubular axoneme with [9(2) + 2] substructure is covered by plasma membrane and limited to eukaryotic cells. Includes cilia and eukaryotic "flagella." Each undulipodium invariably develops from its kinetosome. The principal protein component of microtubules, and thus a major structural protein of the undulipodia, is *tubulin.* Tubulin is a dimer composed of alpha and beta subunits, each of molecular weight 55 kd. See *flagellum* and *microtubule.*

uniformitarianism The fundamental principle or doctrine that geologic processes and natural laws now operating to modify the Earth's crust have acted in the same regular manner and with similar intensity throughout geologic time, and that past geologic events can be explained by phenomena and forces observable today; the concept that the present is the key to the past.

vaterite A rare hexagonal mineral: $CaCO_3$. It is trimorphous with calcite and aragonite, and consists of a relatively unstable form of calcium carbonate. See *limestone.*

Venera A series of Soviet space missions to Venus (1975 to present) emphasizing surface science from survivable "soft landers" (as opposed to the atmospheric science emphasized on the NASA Mariner and Pioneer missions).

Viking NASA exploratory mission to Mars (1975–76) involving two orbiters, astronomical measurements, and two landers replete with scientific instruments. Viking sought life on Mars.

virus An ultramicroscopic, obligate, intracellular, small genome incapable of autonomous replication. Viruses are not autopoietic entities; they reproduce only by entering a host cell and using its protein synthetic system.

Voyager Comprehensive reconnaissance missions, launched in the late 1970s, to Jupiter, Saturn, Uranus, and their satellites; forerunner of Galileo.

xenobiotics Organic comounds of documented external origin.

xenosome An organelle of documented external origin (i.e., kappa particles or *Chlorella* of *Paramecium*).

Addresses of Contributors

Robert Buchsbaum
Massachusetts Audubon: North
Shore
Endicott Regional Center
346 Grapevine Road
Wenham, MA 01984

David Deamer
Department of Chemistry and
Biochemistry
University of California
Santa Cruz, CA 95064

Stjepko Golubic
Department of Biology
Boston University
5 Cummington Street
Boston, MA 02215

Aaron Haselton
Department of Entomology
University of Massachusetts,
Amherst
Amherst, MA 01003

Jonathan King
Department of Biology
68-330
Massachusetts Institute of
Technology
Cambridge, MA 02139

Antonio Lazcano
Faculty of Sciences
Department of Biology
Universidad Autónoma de
México
04511 Mexico D.F.

James Lovelock
Coombe Mill
St. Giles on the Heath
Launceston
Cornwall PL15 9RY
United Kingdom

Lynn Margulis
Department of Geosciences
University of Massachusetts
Amherst, MA 01003

Clifford Matthews
Department of Chemistry
(M/C 111)
University of Illinois
845 West Taylor Street,
Room 4500
Chicago, IL 60607

Michael McElroy
Pierce Hall
Harvard University
Cambridge, MA 02138

Mark McMenamin
Department of Earth and
Environmental Sciences
Mount Holyoke College
South Hadley, MA 01075

Raymond Siever
Department of Geology
Harvard University
Cambridge, MA 02138

Paul Strother
Department of Geology and
Geophysics
Weston Observatory
381 Concord Road
Weston, MA 02193

Neil Todd
Carnivore Genetic Newsletter
26 Walnut Place
Newtonville, MA 02160

Index